The Robotic Process Automation Handbook

A Guide to Implementing RPA Systems

Tom Taulli

Apress®

The Robotic Process Automation Handbook: A Guide to Implementing RPA Systems

Tom Taulli
Monrovia, CA, USA

ISBN-13 (pbk): 978-1-4842-5728-9   ISBN-13 (electronic): 978-1-4842-5729-6
https://doi.org/10.1007/978-1-4842-5729-6

Copyright © 2020 by Tom Taulli

Managing Director, Apress Media LLC: Welmoed Spahr
Acquisitions Editor: Natalie Pao
Development Editor: James Markham
Coordinating Editor: Jessica Vakili

Distributed to the book trade worldwide by Springer Science+Business Media New York, 233 Spring Street, 6th Floor, New York, NY 10013. Phone 1-800-SPRINGER, fax (201) 348-4505, e-mail orders-ny@springer-sbm.com, or visit www.springeronline.com. Apress Media, LLC is a California LLC and the sole member (owner) is Springer Science + Business Media Finance Inc (SSBM Finance Inc). SSBM Finance Inc is a **Delaware** corporation.

For information on translations, please e-mail rights@apress.com, or visit http://www.apress.com/rights-permissions.

Apress titles may be purchased in bulk for academic, corporate, or promotional use. eBook versions and licenses are also available for most titles. For more information, reference our Print and eBook Bulk Sales web page at http://www.apress.com/bulk-sales.

Any source code or other supplementary material referenced by the author in this book is available to readers on GitHub via the book's product page, located at www.apress.com/978-1-4842-5728-9. For more detailed information, please visit http://www.apress.com/source-code.

Printed on acid-free paper

Table of Contents

About the Author

Tom Taulli has been developing software since the 1980s. In college, he started his first company, which focused on the development of e-learning systems. He created other companies as well, including Hypermart.net that was sold to InfoSpace in 1996. Along the way, Tom has written columns for online publications such as BusinessWeek.com, TechWeb.com, and Bloomberg.com. He also writes posts on artificial intelligence (AI) and RPA for Forbes.com and is an advisor to various companies in the space. You can reach Tom on Twitter (@ttaulli) or through his web site (Taulli.com) where he has online courses.

About the Technical Reviewer

John Hindle is a founding partner of Knowledge Capital Partners, a global consultancy offering research, executive education, and advisory services. With a 35-year career as senior executive and advisor in technology industries, he currently serves as vice-chair of the IEEE 2755 Working Group on Intelligent Process Automation. John holds a doctoral degree from Vanderbilt University and publishes extensively in business, trade, and academic media. His most recent book, with Leslie Willcocks and Mary Lacity, is *Becoming Strategic with Robotic Process Automation*.

Foreword

Robotic process automation, or RPA, is the fastest-growing enterprise software segment in history. Much like his last book in this series, *Artificial Intelligence Basics*, Tom Taulli has managed to capture the essence of a highly complex and fast-changing market in thirteen brief chapters. Anyone getting started with RPA would be well served to start here.

RPA's arrival could not have been better timed. RPA products went mainstream just as the world's economy was booming, employers faced the tightest job market in memory, and rising labor costs made business process outsourcing no longer an attractive solution. This confluence of forces set the stage for a tech breakout, the likes of which we haven't seen since Netscape.

The RPA value proposition is seductively simple – a tireless army of software robots (bots) that will work night and day to tackle the mountain of labor-intensive data entry work that sustains our digital world. And RPA should be fast to implement – no more waiting months for expensive APIs to be developed for legacy integration since bots interface with your systems by logging in just like humans do. Watch a bot cut and paste a few hundred fields for you in a split second, and it's hard not to get a little excited.

In three short years, the current leading vendors in the stand-alone RPA market, UiPath, Automation Anywhere, and Blue Prism, have defied all the cynics and grown from small start-ups to multibillion dollar valuations, thousands of employees and corporate clients, and they continue to post triple-digit revenue growth. They have also left behind mixed business results, layoffs, and questions about what use cases are actually best suited to RPA and which are better served with other low-code automation technologies.

To cut through the hype, Tom has interviewed leaders and expert practitioners from across the RPA space and brought together their collective wisdom and real-world experience in this single volume. The result is a practical guide to successfully deploying RPA in your business. As a technology partner to many of the pure-play RPA vendors covered in this book, I found Tom's treatment of the market useful and objective.

In these pages, you will progress rapidly from selecting the right use case for RPA vs. other automation methods to scaling your RPA program. You'll learn how RPA integrates with technologies like AI and low-code automation to provide complete solutions. You'll also read examples of how companies are establishing new organizational models to govern and manage a successful, enterprise-scale robotic workforce.

As with any revolutionary new technology, starting with RPA has proven relatively simple but scaling RPA projects is proving more challenging. The result is enterprise deployment timelines sometimes extend to twenty-four months and higher than normal failure rates reported by industry analysts. Experienced leadership and best practices are hard to come by because the market and its participants are so new and immature, and the demand for their expertise is explosive.

Exaggerated expectations and marketing hype make up a substantial portion of the problem Tom's book addresses. RPA is "robotic" automation, but these aren't robots; they can't sense and respond to change anything as humans do. In the right applications and hands, RPA is a powerful tool to augment a human workforce, transform the customer experience, and accelerate digital transformation. RPA programs require planning and governance to ensure they continue to deliver the results you expect.

Where does the RPA market go from here? We can see the future in how the largest adopters of RPA are increasingly viewing it as part of an overall automation program that includes human workflow, API integration, and AI in a unified platform. Digital transformation is only truly enabled when organizations break down the process and data silos to orchestrate work among digital workers, humans, and algorithms. RPA is also moving off the desktop into the cloud for better security, governance, efficiency, and scalability – top CIO priorities.

Customer demand for full-stack automation can be seen in both the addition of basic human workflow capabilities and cloud options being introduced by RPA vendors. It is also happening by acquisitions – like PEGA's purchase of attended automation specialist OpenSpan, covered in this book, and in Appian's recently announced acquisition of Jidoka RPA, the top-ranked Gartner Peer Insights RPA product. While too recent to be included in this book, Appian's addition of RPA to its existing Robotic Workforce Manager epitomizes the convergence we are witnessing between the RPA and low-code automation markets as well as the migration of RPA from the desktop to the cloud.

This is an exciting time to dive into the RPA market. This book will help you chart a path to success with RPA and avoid the pitfalls that have tripped up others on their automation journey.

Michael Beckley
Founder and Chief Technology Officer, Appian

Introduction

In Silicon Valley today, one of the buzziest categories is RPA, or robotic process automation, as venture capitalists have been investing huge sums in the category. The reason is simple: the software providers have been ramping at a staggering rate.

In fact, according to a recent survey from LinkedIn, RPA is the second fastest growing career category, up 40% in 2019. The report notes: "Careers in Robotics Engineering can vary greatly between software and hardware roles, and our data shows engineers working on both virtual and physical bots are on the rise."[1]

But as with any rapidly growing industry – especially in the technology sector – there is lots of misinformation and hype. It's inevitable.

Yet this can make it difficult for companies to evaluate the technology. What are the best solutions? What are the success factors for implementation? What are the gotchas?

Well, in this book, we'll help to answer these questions – and many more. Yes, it is your handbook for navigating the noise in the RPA marketplace.

There are certainly many factors that help explain the surge in RPA. But perhaps the most important is the need for digital transformation. Simply put, companies do not want to be flat-footed when a disruptive start-up upends an industry, as seen with Uber.

---

[1]https://business.linkedin.com/content/dam/me/business/en-us/talent-solutions/emerging-jobs-report/Emerging_Jobs_Report_U.S._FINAL.pdf

As former Cisco CEO John Chambers has written in his book *Connecting the Dots*: "But I also now understand the fears because this disruption will be so brutal that 40-plus percent of businesses today won't be here 10 years from now."

However, if companies take the initiative to craft the right strategies and carry them out, the future can be bright, even for those that are in traditional industries. If anything, they have certain advantages that can be leveraged: trusted brands, extensive distribution and customer bases, and talented employees.

No doubt, automation will be critical. "The disparity between automated and non-automated companies will grow," said Prince Kohli, who is the CTO of Automation Anywhere. "Research has shown more than half of businesses in North America have already implemented some type of automation solution – meaning that we have now reached the tipping point of adoption for this transformative technology. Given the dramatic increases in productivity, reductions in cost and improvements to employee satisfaction that RPA provides, we'll soon see the gap widen between early adopters and holdouts – in terms of revenue, customers, talent acquisition and retention – making it much harder for the latter to compete."[2]

OK then, so what will we cover in this book? Let's take a look at a quick outline:

Chapter 1 – RPA Foundations: This is a high-level overview of the RPA industry, with coverage of assisted/unassisted automation, the history of the technology, the benefits and drawbacks, and comparisons to other automation technologies.

Chapter 2 – RPA Skills: This chapter takes a look at core technologies like on-premise software, cloud computing, OCR (optical character recognition), databases, APIs (application programming interfaces), and AI. There are also explanations of programming techniques, such as Agile and DevOps. Finally, there is a tutorial on flowcharts, which are crucial for RPA.

---

[2]From the author's interview with Prince Kohli, the CTO of Automation Anywhere.

Chapter 3 – Process Methodologies: Before implementing RPA, a company may want to ensure that its processes are in good shape. This can be done by using methodologies like lean, Six Sigma, and lean Six Sigma.

Chapter 4 – Planning: This chapter provides help on assessing processes, determining what to automate, and how to measure ROI. We also show how to hire an RPA consultant.

Chapter 5 – RPA Vendor Evaluation: Here you will find the steps in selecting the right RPA software, such as looking at costs, training, functionality, and security.

Chapter 6 – Center of Excellence (CoE): This is a group that helps manage an RPA implementation. So in this chapter, we'll show how to assemble one and how to best utilize it. We also take a look at the main roles for RPA (business analysts, developers, managers, etc.).

Chapter 7 – Bot Development: You'll get a fundamental understanding of how to create a bot, such as developing the structure, using variables, structuring workflows, and debugging. The chapter focuses on the UiPath software system.

Chapter 8 – Deployment and Monitoring: Here you will see how to put bots into production and make sure they are functioning properly. Some best practices on scaling are also explained.

Chapter 9 – Data Preparation: RPA can be a great way to transition to AI. But to do this, there needs to be a strong data strategy. This chapter shows a technique for this, called the CRISP-DM Process.

Chapter 10 – RPA Vendors: This is a review of not only the large RPA software developers, like UiPath, Automation Anywhere, and Blue Prism, but also the smaller ones.

Chapter 11 – Open Source RPA: This category of RPA is small but it is poised for growth. This chapter will highlight some of the more notable open source projects.

Chapter 12 – Process Mining: What started as an academic area of research has recently turned into a large industry. The technology helps to map and optimize processes – and has become increasingly important to RPA.

Chapter 13 – Future of RPA: Here we look at some of the major trends in the industry, like the changing of the business model, the growth of cloud systems, and the importance of AI.

At the back of the book, you'll also find an appendix of resources for further study, a glossary of common RPA terms, and a list of RPA consultants.

Accompanying Material

Any updates will be provided on my site at `www.Taulli.com`.

CHAPTER 1

RPA Foundations

What the Technology Can Do

RPA (robotic process automation) has become one of the hottest categories for venture capital investment. In November 2018, Automation Anywhere announced that the Softbank Vision Fund invested $300 million in the start-up.[1] But it was not the only amount. The Series A round was actually over $500 million and included strategic investors like Workday, which is a top cloud-based ERP (enterprise resource planning) operator. The managing director and cohead of the Workday Ventures arm noted: "RPA is becoming a bigger focus for our customers, which is why it's important for us to partner closely with Automation Anywhere, a market leader."[2]

Then there was another mega round for a fast-growing RPA vendor. In April 2019, UiPath announced it raised $568 million for its Series B round, with tier-one investors like Coatue, Wellington, CapitalG, Accel, and Sequoia.

In only about two years, UiPath had seen an explosion in growth:

- The valuation went from $110 million to $7 billion.

- The annual recurring revenues surged from $8 million to $200 million.

- The employee base jumped by 16X to 2,500.

---

[1] www.automationanywhere.com/uk/press-room-uk/automation-anywhere-announces-%24300-million-investment-from-the-softbank-vision-fund

[2] www.prnewswire.com/news-releases/automation-anywhere-announces-strategic-investment-from-workday-ventures-300791701.html

© Tom Taulli 2020
T. Taulli, *The Robotic Process Automation Handbook*,
https://doi.org/10.1007/978-1-4842-5729-6_1

- There was also the launch of six new releases of the
  UiPath Enterprise RPA platform.

- The company snagged many large customers like
  McDonald's, Duracell, Google, Nippon Life Insurance,
  Ricoh, and Voya Financial.

The CEO and cofounder of UiPath, Daniel Dines, said: "We are
at the tipping point. Business leaders everywhere are augmenting
their workforces with software robots, rapidly accelerating the digital
transformation of their entire business and freeing employees to spend
time on more impactful work. UiPath is leading this workforce revolution,
driven by our core determination to democratize RPA and deliver on our
vision of a robot helping every person. I am humbled by the amazing
support our customers, partners and investors give us every day, inspiring
us to work harder to evolve RPA as the platform that not only unlocks the
true potential of AI, but also other emerging technologies. We are just
getting started."[3]

Yes, it's exciting times for the RPA industry and there are few signs of
a slowdown, at least in terms of customer demand. So then why all the
interest? What are some of the key drivers for RPA? And besides, what
really is RPA?

In this chapter, we'll answer these questions as well as get a foundation
on the core elements of the technology.

What Is RPA?

RPA can be a slippery term. A main reason for this is that it was coined in
2012, when the category was still evolving. At the time, the chief evangelist
for Blue Prism, Pat Geary, came up with the term RPA.

---

[3]www.uipath.com/newsroom/uipath-raises-568-million-series-d-funding-round

But when you look at each word, it's understandable why RPA can be a bit confusing. For example, the word "robotic" does not refer to a physical robot – instead, it is about a software-based robot (or bot) that can automate human actions in the workplace (generally for white collar applications in clerical and administrative functions). A bot can be delivered via the cloud or through downloadable software. However, the use of robotic does look like a savvy marketing move (hey, aren't robots pretty cool?).

Even the word "process" is not particularly descriptive either. A better alternative would be "tasks," which are individual action items that are a part of a process.

OK, then what really is RPA? Well, in a nutshell, RPA involves bots that perform a set of specified actions or tasks, such as the following:

- The cut-and-paste of information from one app to another

- The opening of a web site and login

- The opening of an e-mail and attachments

- The read/write of a database

- The extraction of content from forms or documents

- The use of calculations and workflows

Such things may sound kind of mundane, boring, and simplistic. But that's the point. RPA is focused on those tasks that are really a waste of efforts for workers. Shouldn't they be doing more important activities?

I think so.

Now, interestingly enough, the use of the word "automation" in RPA is actually spot-on. It's really at the core of RPA functionality.

To get a better sense of all this, I think it's a good idea to look at how various RPA software companies view the concept. Here's a look:

UiPath: "Robotic Process Automation is the technology that allows anyone today to configure computer software, or a 'robot' to emulate and integrate the actions of a human interacting within digital systems to execute a business process. RPA robots utilize the user interface to capture data and manipulate applications just like humans do. They interpret, trigger responses and communicate with other systems in order to perform on a vast variety of repetitive tasks. Only substantially better: an RPA software robot never sleeps, makes zero mistakes and costs a lot less than an employee."[4]

Automation Anywhere: "RPA is really as simple – and powerful – as it sounds. Robotic Process Automation enables you with tools to create your own software robots to automate any business process. Your 'bots' are configurable software set up to perform the tasks you assign and control.

"Think of them as your Digital Workforce. Show your bots what to do, then let them do the work. They can interact with any system or application the same way you do. Bots can learn. They can also be cloned. See how they are working and adjust and scale as you see fit. It's code-free, non-disruptive, non-invasive, and easy."[5]

PEGA: "Robotic process automation (RPA) can be a fast, low-risk starting point for automating processes that rely on outdated legacy systems. Bots can pull data from manual systems without APIs into digital processes, ensuring faster and more efficient outcomes.

"Now, let's be honest about what RPA doesn't do. It doesn't transform your organization all by itself, and it's not a fix for enterprise-wide broken processes and systems. For that, you'll need end-to-end intelligent automation."[6]

Kryon Systems: "Robotic Process Automation enables enterprises to create true virtual workforces that drive business agility and efficiency. A virtual workforce, comprised of software robots that can execute business tasks on enterprise applications, becomes an integral part of an

---

[4]www.uipath.com/rpa/robotic-process-automation
[5]www.automationanywhere.com/robotic-process-automation
[6]www.pega.com/rpa

enterprise's greater workforce. It is managed just as any other team in the organization and can interact with people just as other employees would interact with one another. Virtual workers (robots) complete business processes, just as a person would, but in less time, with greater accuracy and at a fraction of the cost. RPA stands out for its ability to impact business outcomes, resulting in significant ROI."[7]

These all provide a fairly good view of RPA, showing the broad applications and benefits. These definitions also highlight that the vendors in the industry have their unique twists and approaches on the technology (in this book, we'll take a deeper look at the different solutions). In some cases, the differences can be quite stark.

But boiling things down, I think the best way to think of RPA is to use the visual for Automation Anywhere: a digital worker. It's about how automation technologies – like screen scraping and workflows – can essentially copy what employees do on a daily basis.

Note   Keep in mind that it's not uncommon for companies to name their bots!

But what is the difference with RPA vs. other forms of automation? Isn't the technology just like an Excel macro?

Not really. First of all, a macro is only for a particular application. But with RPA, the system can be used for just about anything, whether on a PC or even a mainframe. Next, RPA can record a person's actions to help create the automation. Some systems will even use sophisticated technologies like AI (Artificial Intelligence) to help with this. Finally, an RPA platform will become a valuable repository of knowledge about how work is done in an organization. This can provide insights on how to improve workflows and processes, which could drive even further efficiency.

---

[7]www.kryonsystems.com/what-is-rpa/

Flavors of RPA

There are different types of RPA approaches. Part of this is due to the fact that the technology is continuing to evolve. Vendors are also looking at ways to redefine RPA so as to help them stand out in the marketplace.

On a high level, you can divide the flavors into the following:

- Attended RPA (which may be referred to as robotic desktop automation or RDA): This was the first form of RPA that emerged, back in 2003 or so. Attended RPA means that the software provides collaboration with a person for certain tasks. A prime example would be in the call center, where a rep can have the RPA system handle looking up information while he or she talks to a customer.

- Unattended RPA: This technology was the second generation of RPA. With unattended RPA, you can automate a process without the need for human involvement – that is, the bot is triggered when certain events happen, such as when a customer e-mails an invoice. Consider that unattended RPA is generally for back-office functions.

- Intelligent process automation or IPA (this may also be referred to as cognitive RPA): This is the latest generation of RPA technology, which leverages AI to allow the system to learn over time (an example would be the interpretation of documents, such as invoices). Because of this, there may be even less human intervention since the RPA software will use its own insights and judgements to make decisions.

It's important to understand these variations because some RPA systems may specialize in a particular approach. Besides, when looking at your own needs for automation, it's a good idea to see what types may work the best.

History of RPA

No doubt, the concept of automation is far from new. Did you know that the first mention of the concept was from Homer's *The Iliad*? In the poem, he described how Hephaestus (the Greek god of blacksmiths) used automatons (or machines) to build weapons for the gods of Mount Olympus.

Yet it would not be centuries later until notable real-world examples of automation would emerge. After all, it's only been during the past 70 or so years that computers have been a major catalyst for this trend.

Along the way, there have been different periods of automation, based on the types of technologies available. They would also provide a foundation for RPA platforms.

- Mainframe Era: These were huge machines developed by companies like IBM. They were expensive and mostly available to large companies (although, innovators like Ross Perot would create outsourcing services to provide affordable options). Yet they were incredibly useful in helping manage core functions for companies, such as payroll and customer accounts.

- PC Revolution: Intel's development of the microprocessor and Microsoft's development of its operating system revolutionized the technology industry. As a result, just about any business could automate processes, say by using word processors and spreadsheets.

But the automation technologies – while powerful – still had their drawbacks. They could easily result in complex IT environments, which required expensive and time-consuming integrations and custom coding. Because of this, an employee may have to use multiple applications in their daily activities that could involve wasteful tasks like moving data from one to the other. The irony was that the technology could make employees less productive!

From this emerged the key elements for RPA, which came about in the early 2000s. A big part of this was screen scraping, which is the automation of moving data among applications, which turned out to provide a nice boost to efficiency and effectiveness.

But the nascent RPA market got scant attention. It was mostly perceived as low-tech and a commodity. Instead, investors and entrepreneurs in Silicon Valley focused their attention on the rapidly growing cloud market that was disrupting traditional IT systems.

But around 2012 or so, the RPA market hit an inflection point. There was a convergence of trends that made this happen, such as the following:

- In the aftermath of the financial crisis, companies were looking for ways to lower their costs. Simply put, traditional technologies like ERP were reaching maturation. So companies needed to look for new drivers.

- Companies also realized they had to find ways to not be disrupted from technology companies. RPA was considered an easier and more cost-effective way to go digital.

- Some industries like banking were becoming more subject to regulation. In other words, there was a compelling need to find ways to lessen the paperwork and improve audit, security, and control.

- RPA technology was starting to get more sophisticated and easier to use, allowing for higher ROI (return on investment).

- Large companies were starting to use RPA for mission-critical applications.

- Demographics were also key. As the millennials started to enter the workforce, they wanted more engaging work. They wanted careers, not jobs.

"The evolution of the RPA market is like any major technology trend," said Mihir Shukla, who is the CEO and cofounder of Automation Anywhere. "There was a gradual progress, which involved periodic breakthroughs. A prime example is the iPhone. Before this, there was a long period of incremental innovation."[8]

Fast forward to today, RPA is the fastest growing part of the software industry. According to Gartner, the spending on this technology jumped by 63% to $850 million in 2018 and is forecasted to reach $1.3 billion by 2019.[9]

Or consider the findings from Transparency Market Research. The firm projects that the global market for RPA will soar to $5 billion by 2020.[10]

Here are some other metrics to note:

- By 2020, RPA along with AI will reduce the business shared-service centers by 65% (Gartner). There will also be adoption by 40% of large enterprises, compared to 10% in 2019.[11]

---

[8]From the author's interview with Automation Anywhere CEO Mihir Shukla (on October 9, 2019).

[9]www.idgconnect.com/analysis-review/1502790/robotic-process-automation-trend-enterprise-digitalisation

[10]www.cio.com/article/3124638/why-bots-are-poised-to-disrupt-the-enterprise.html#tk.cio_rs

[11]www.cio.com/article/3236451/what-is-rpa-robotic-process-automation-explained.html

- Based on current projections, there will likely be saturation in the RPA market by 2023 (Deloitte).[12]

- The financial impact from RPA could hit $6.7 trillion by 2025 (McKinsey & Company).[13]

- In terms of the global market share for RPA software, North America represents 51% and Western Europe is at 23%. But Asia is starting to get traction, especially Japan.[14]

- By 2023, the forecast is that there will be $12 billion in spending on RPA services (Forrester).[15]

The Benefits of RPA

When it comes to RPA, the most talked about benefit is the ROI. Compared to just about any other enterprise software technology, the metrics are standout. Take the Computer Economics Technology Trends 2019 report, which is a survey of 250 companies (the study covered many industries that had revenues from $20 million to billions). Among them, about 12% implemented RPA within their organizations and half of them said there was a positive ROI within 18 months (the remaining was mostly at breakeven).[16] "We expect RPA to grow rapidly, because of the

---

[12]www2.deloitte.com/ro/en/pages/technology-media-and-telecommunications/articles/deloitte-global-rpa-survey.html

[13]https://globalpayrollassociation.com/blogs/technology/what-the-history-of-rpa-technology-says-about-its-future

[14]www.gartner.com/en/newsroom/press-releases/2019-06-24-gartner-says-worldwide-robotic-process-automation-sof

[15]www.forrester.com/report/The+RPA+Services+Market+Will+Grow+To+Reach+12+Billion+By+2023/-/E-RES156255#

[16]www.wsj.com/articles/unleash-the-bots-firms-report-positive-returns-with-rpa-11551913920

success of early adopters," said David Wagner, who is the vice president for research at Computer Economics.[17]

Here's another perspective from consulting firm, A.T. Kearney: "On average, a software robot costs one-third as much as an offshore employee and one-fifth as much as onshore staff. Several prominent service industry firms have seen cost reduction and process improvement from the use of robots. Barclays Bank attributes savings worth the equivalent of roughly 120 full-time employees and an annual reduction in bad debt provisions of $250 million. Telefónica O2, which uses more than 160 robots to automate 15 core processes and nearly 500,000 transactions per month, says that its return on investment in robotic process automation has exceeded 650 percent."[18]

But when looking at RPA, the benefits are far more than just about the impact on the bottom line. The technology can transform a company.

So let's take a look:

The Impact of Small Improvements: On the surface, an employee who saves 10 to 20 seconds on a task – even something as simple as a series of cut-and-paste actions – may seem trivial. But it's not. When scaled across thousands of employees across a global organization, the impact can certainly be significant. For example, some companies will keep track of the metric of how many hours are saved by using RPA, which becomes a part of the overall ROI calculation.

Note   A survey form Forrester found that 86% of the respondents reported an increase in efficiency from RPA.[19]

---

[17]www.computereconomics.com/article.cfm?id=2633

[18]www.atkearney.com/documents/20152/969206/Robotic+Process+Automation.pdf/49de900b-8646-5636-6fd4-b79be8d54e43?t=1515625008538

[19]www.cio.com/article/3433181/the-dark-side-of-robotic-process-automation.html

Relative Ease of Implementation: Unlike traditional business applications like a CRM or ERP, RPA generally does not involve an onerous implementation and integration. Why? Note that the software sits on top of existing IT systems. RPA is also relatively easy for a person to use since there is no requirement for understanding complex coding. As a result, there is not as much reliance on the IT department for support, which is certainly a win–win, or a need for heavy training. The bottom line: The people implementing RPA will get to their objectives quicker and the IT department will have more time to devote to higher priority items. This is important as there remains a trend of less investment in IT.

Compliance: Just one violation of a government regulation can have a serious adverse impact on a company. It could even be a threat to its very existence, as we have seen with examples like Enron or Theranos.

While employees are usually diligent and trustworthy, they do make mistakes or they may not understand some of the regulations. Yet this is not an issue with RPA. You can easily configure a bot to make sure the actions are compliant with regulatory requirements. Many RPA vendors also have built in their own compliance systems, handling such laws as the Sarbanes–Oxley Act, General Data Protection Regulation (GDPR), and HIPAA (Health Insurance Portability and Accountability Act of 1996).

Note   In a survey form NICE, the respondents indicated that compliance was where there was the highest level of improvement with RPA software.[20]

Another compliance benefit is that there will be less intervention with the data from people, which lessens the possibility of fraud. What's more, RPA provides a strong audit trail to allow for better tracking and monitoring.

---

[20]https://enterprisersproject.com/article/2019/9/rpa-robotic-process-automation-14-stats?page=1

Customer Service: Nowadays, people want quick and accurate responses from their companies. But this is difficult to provide, especially when a company is overwhelmed from incoming contacts.

But this is where RPA can make a big difference. The bots are programmed to make sure that all the necessary steps are taken – at scale. The result is often an increase in customer satisfaction metrics, like the Net Promoter Score (NPS).

"RPA can take the processing of a mortgage application from 15 days to 7 minutes," said Shukla. "It's something like this that can go a long way with the customer experience."[21]

Employee Satisfaction: Yes, your team should also enjoy the benefits of RPA. After all, it means that they do not have to spend their valuable time on tedious activities. The result may be less turnover and higher productivity.

Note   A survey from Forrester found that RPA increased employee engagement for 72% of the participants.[22]

Wide Application: It's common for an enterprise application to focus on a certain part of a company's departments or functions. But RPA is wide. It can be used for virtually any part of a company, such as legal, finance, HR, marketing, sales and so on.

Data Quality: It should be greatly improved as there will be less chance of human error. In fact, there will probably be much more data because of the scalability of the automation. In other words, the datasets for analytics and AI will be more robust and useful.

---

[21]From the author's interview with Automation Anywhere CEO Mihir Shukla (on October 9, 2019).

[22]www.automationanywhere.com/blog/changing-the-world-with-automation/impress-your-employees-with-rpa

Digital Transformation: This is a major priority for CEOs. But many companies have legacy systems that would be expensive to replace or integrate. However, RPA is an approach that can help with this process, which is often quicker and less costly.

Scalability: If there is a sudden jump in demand, it can be extremely difficult to hire new employees. But RPA can be a solution. It is much cheaper and faster to ramp up new bots to meet the demand.

The Downsides of RPA

RPA is definitely not a cure-all. The software has its inherent limitations and complexities.

So here's a look:

> Cost of Ownership: The business models vary. Some have a subscription or multiyear license. Other vendors may charge based on the number of bots.

> But there is more to the costs. There is the need for some level of training and ongoing maintenance. Depending on the circumstances, there may be requirements for buying other types of software and hardware. Oh, and it is common to retain third-party consultants to help with the implementation process.

> Technical Debt: This describes software that is not a comprehensive solution that ultimately requires ongoing reworking, updates, and changes. And yes, this is an issue with RPA. As a company's processes change, the bots may not work properly. This is why RPA does require ongoing attention.

Enterprise Scale: This is when RPA is pervasive across the whole organization. While this can result in major benefits, there are also potential land mines. It can be extremely difficult to manage the numerous bots and there also needs to be strong collaboration among IT.

Note   A study from Deloitte UK found that only 3% of organizations were able to scale RPA to 50 or more bots.[23]

Security: This is a growing risk with RPA implementations, especially as the technology covers more mission-critical areas of a company's processes. Let's face it, if there is a breach, then highly sensitive information could easily be obtained. Actually as RPA gets more pervasive in manufacturing, there may even be risks of property damage and bodily harm. This would likely be the case with attended RPA.

Expectations: With the hype at feverish levels for RPA (it's a top headline grabber for many business and technology publications), this could easily lead to disappointment. According to a survey from PEGA, the average time it takes to develop a quality bot was 18 months, with only 39% being deployed on time.[24]

---

[23]www.cio.com/article/3124638/why-bots-are-poised-to-disrupt-the-enterprise.html#tk.cio_rs
[24]www.pega.com/about/news/press-releases/survey-most-businesses-find-rpa-effective-hard-deploy-and-maintain

Preparation: You need to do a deep dive in how your current tasks work. If not, you may be automating bad approaches! In the next chapter, we'll take a look at the best practices for avoiding this problem.

Limits: RPA technology is somewhat constrained. For the most part, it works primarily for tasks that are routine and repetitive. If there is a need for judgment – say to approve a payment or to verify a document – then there should be human intervention. Although, as AI gets more pervasive, the issues are likely to fade away. For example, insurance companies can use the technology to adjudicate claims for payments, based on individual claims history and firm-wide payment policies.

Virtualized Environments: This is where a desktop accesses applications remotely, such as through a platform like Citrix. Yet this can make an RPA system fail. How? The reason is that it cannot capture the text on the screen. However, some of the latest RPA offerings, such as from UiPath, are solving the problem.

RPA Compared to BPO, BPM, and BPA

In the discussion about RPA, you may hear terms like business process management (BPM), business process outsourcing (BPO), and business process automation (BPA). They can get kind of confusing but they have key distinctions.

Here's a look:

BPM: With the intense competition from Japan during the 1970s and 1980s, US companies were desperately seeking new and innovative approaches to improve their efficiency and competitiveness. Part of this meant adopting different management approaches, such as Six Sigma (this includes a combination of project management and statistical techniques), lean production (which is based on the manufacturing principles of Toyota), and total quality management or TQM (a blend of Six Sigma and lean production). There was also a greater focus on computer technologies. For example, FileNet introduced a digital workflow management system to help better handle documents (the company would eventually be purchased by IBM). Then there would come onto the scene ERP vendors, such as PeopleSoft.

All of this would converge into a major wave called BPM. For the most part, the focus was on having a comprehensive improvement on business processes. This would encompass both optimizing systems for employees but also IT assets.

There were also various business process management software (BPMS) solutions to help implement BPM. One was Laserfiche. Nien-Ling Wacker founded the company in 1987, when she saw the opportunity to use OCR (optical character recognition) technology to allow users to search huge volumes of text.

So then how is BPM different from RPA? With BPM, it requires much more time and effort with the implementation because it is about changing extensive processes, not tasks. There also needs to be detailed documentation and training. Because of this rigorous approach, BPM is often attractive to industries that are heavily regulated, such as financial services and healthcare. However, the risk is that there may be too much structure, which can stifle innovation and agility.

On the other hand, RPA can be complementary to BPM. That is, you can first undergo a BPM implementation to greatly improve core processes. Then you can look to RPA to fill in the gaps.

Here's how a blog from UiPath describes things: "Consider this analogy to self-driving cars: a BPM approach would require us to rip up all paved roads and install infrastructure for the new cars to move about on their own, while an RPA approach seeks to operate a pre-existing car just as a human would. Google has come at the problem from an RPA angle, because replacing all roads (especially in the U.S.) is just unfathomable. That's not to say that RPA is always the better option – not at all. The key is knowing the difference and using both tactics to their best advantage."[25]

BPO: This is when a company outsources a business service function like payroll, customer support, procurement, and HR. The market is massive, with revenues forecasted to reach $343.2 billion by 2025 (according to Grand View Research).[26] Some of the top players in the industry include ADP, Accenture, Infosys, IBM, TCS, and Cognizant.

As should be no surprise, one of the big attractions of BPO is the benefit of lower wage rates in other countries (this is often referred to as "labor arbitrage"). The employee bases will also often be educated and multilingual.

But there are certainly other major advantages. For example, a company does not have to waste its attention on noncore functions. In fact, by outsourcing various areas of a company, there is the benefit of having a specialist provide the service, which should mean getting better results.

Generally speaking, a BPO will have three types of strategies:

- Offshore: This is where the employees are in another country, usually far away.

---

[25]www.uipath.com/blog/rpa-vs-bpm-one-goal-two-solutions

[26]www.prnewswire.com/news-releases/business-process-outsourcing-market-worth-343-2-billion-by-2025-grand-view-research-inc-300794746.html

- Nearshore: This is when the BPO is in a neighboring country. True, there are usually higher costs but there is the benefit of being able to conveniently visit the vendor. This can greatly help with the collaboration.

- Onshore: The vendor is in the same country. For example, there can be wide differences in wages in the United States.

But as with anything, there are drawbacks with a BPO. Perhaps the most notable one is the quality issue (you know the situation when you call a company and get an agent you can barely understand!).

Yet here are some others to consider:

- Security: If a BPO company is developing an app with your company's data, are there enough precautions in place so there is not a breach? Even if so, it can still be difficult to enforce and manage.

- Costs: Over the years, countries like China and India have seen rising labor costs. This has resulted in companies moving to other locations, which can be disruptive and expensive.

- Politics: This can be a wildcard. Instability can easily mean having to abandon a BPO operator in a particular country.

Now as for RPA, what is the connection then with BPO? There is an interesting one. RPA is automating BPO-type activities. Based on studies from firms like Everest Group, KPMG, and Deloitte, the cost advantages of RPA over outsourcing can be as much as 70%.[27] This may mean there will

---

[27]https://blog.sageintacct.com/blog/will-robotic-process-automation-rpa-replace-business-process-outsourcing

be less outsourcing in the coming years. Yet BPO companies have been adapting to this. Keep in mind that these firms are generally heavy users of RPA technology.

BPA: This is the use of technology to automate a complete process. One common use case is onboarding. For example, bringing on a new employee involves many steps, which are repeatable and entail lots of paperwork. For a large organization, the process can be time-consuming and expensive. But BPA can streamline everything, allowing for the onboarding at scale.

OK, this kind of sounds like RPA, right? Yes, this is true. But there is a difference in degree. RPA is really about automating a part of the process, whereas BPA will take on all the steps.

Consumer Willingness for Automation

The automation of consumer-facing activities, such as with chatbots on a smartphone or web site, are becoming more ubiquitous. But this begs the question: Does this technology really provide a good experience? Might it be doing more harm than good?

These are very good questions to ask as many automation technologies are far from perfect, especially those that deal with complex environments.

Consider a report from Helpshift, an AI-based digital customer service platform automating 80% of customer support issues for huge D2C (direct-to-consumer) brands including companies like Flipboard, Microsoft, Tradesy, and 60 others. Its report is based on the analysis of 75 million customer service tickets and 71 million bot-sent messages.

Here are some of the findings:

- A total of 55% of the respondents – and 65% of millennials – prefer chatbots with customer service so long as it is more efficient and reduces phone time to resolve an issue and explain a problem.

- A total of 49% say they appreciate the 24/7 availability of chatbots.

Granted, there is much progress to be made. Chatbot technology is still in the early phases and can be glitchy, if not downright annoying in certain circumstances. But in the years to come, this form of automation will likely become more important – and also a part of the RPA roadmap.

According to the CEO of Helpshift, Linda Crawford: "Seeing as the vast majority of Americans dread contacting customer support, there's a huge opportunity here for chatbots to fill the void and improve the customer support experience for consumers—and agents."[28]

The Workforce of the Future

The modern concept of work goes back to Henry Ford. To scale his operations to create the automotive industry, he realized he had to be competitive with wages so as to attract the best employees. He made the remarkable move of doubling them $5 a day, which actually had the consequence of greatly expanding the middle class! But he also introduced other policies, such as the 40-hour workweek, with weekends off.

The interesting thing is that the fundamentals of work have not changed much since then. True, there has been the trend of the gig economy, in which people get paid for offering services through Uber and Lyft. Yet when it comes to office work, the structure has remained quite durable.

---

[28]www.helpshift.com/press/helpshift-study-americans-love-chatbots-get-humans-faster/

But the fact remains that many of the activities remain fairly tedious and uninteresting. According to research from the McKinsey Global Institute, white collar workers still spend 60% of their time on manual tasks, such as with answering e-mails, using spreadsheets, writing notes, and making calls.[29] Interestingly enough, a report from David White estimates that $575 billion is wasted in the United States because of inefficient processes.[30]

In light of all this, RPA is likely to have a significant impact on the workplace because more and more of the repetitive processes will be automated away. One potential consequence is that there may be growing job losses. A survey from Forrester predicts that – as of 2025 – software automation will mean the loss of 9% of the world's jobs or 230 million. Then again, the new technologies and approaches will open up many new opportunities. Yet this may not be enough to make up for the shortfall. The Forrester study, for example, forecasts that there will be replacement of 16% of US jobs and the creation of 9% of new ones.[31]

Or look at the research from McKinsey & Company. Its analysis shows that technologies like RPA could automate a whopping 45% of the activities of a company's workforce.[32]

Now when a company engages in an automation project, the CEO will usually not talk about job loss. It's something that will frighten the workforce and generate awful headlines. Instead the messaging will be vague, focusing on the overall benefits of the transformation.

This may make it sound like not much is happening. But it does seem like a good bet that the reverberations will grow and grow, as RPA systems

---

[29]www.sec.gov/Archives/edgar/data/1366561/000162828018005173/smartsheet424.htm#s2DC3CF561C88FF32AA45F92C369BC64E

[30]www.enterpriseappstoday.com/management-software/3-automation-initiatives-to-boost-corporate-productivity.html

[31]www.cio.com/article/3124638/why-bots-are-poised-to-disrupt-the-enterprise.html#tk.cio_rs

[32]www.uipath.com/blog/automation-first-mindset

get increasingly robust. As we've seen in prior periods where technology resulted in job loss – such as in the Industrial Revolution – there are serious changes in politics and regulations. The upshot is that technology often becomes a target of scorn.

Companies really do try to avoid layoffs, since they are expensive and take a toll on the organization. But in the years ahead, managers will probably need to find ways to navigate the changes from automation, such as finding new roles or reskilling the workforce.

All in all, the rise of automation has the potential for leading for a much better society. Again, workers can focus on more interesting and engaging activities – not repetitive and mundane tasks. There will also be ongoing renewing of knowledge and understanding. But there must be proactive efforts, say from companies and governments, to provide for a smoother transition.

Note   A survey from PEGA shows that nearly 70% of the respondents believe that the concept of "workforce" will include employees and machines and 88% believe this is fine, so long as they are not managed by machines.[33]

Conclusion

As you can see, RPA is really an exciting area of technology. The software has been shown to result in quick ROI and has helped companies – in many industries – to transform themselves. But there are other benefits like improved compliance, better customer service, and greater data quality.

---

[33]www.pega.com/system/files/resources/2018-12/Future-of-Work-Report.pdf

On the other hand, RPA does have its challenges and issues. The technology can be tough to manage, say when there are a large number of bots. There are also potential complications with security and scalability across the enterprise.

Yet despite all this, RPA appears to be a core technology that will be around for the long haul.

Key Takeaways

- RPA is software that helps to automate certain business tasks, such as working with applications like CRM or ERP. One way to think of this is as a digital worker.

- RPA is not a macro. Some of the reasons include the following: RPA can handle many applications and technology environments and there are ways to easily create the automations (such as by recording actions of a worker). There are also other potential features like AI.

- Attended RPA is where the software helps a person with certain tasks. A typical use case is in the call center, where the software can help the agent handle a customer question.

- Unattended RPA is software that runs without human intervention. Usually, the bot takes action when there is a trigger, such as when an invoice is sent via e-mail to the company.

- IPA involves the use of AI and other sophisticated technologies to deal with judgements and better decision-making in the automation process.

- BPM software invovles automation but is much more comprehensive than RPA.

- BPO is when a company outsources a business service function like payroll, customer support, procurement, and HR.

- BPA involves the use of technology to automate a complete process.

CHAPTER 2

RPA Skills

The Technologies You Need to Know

While RPA does not require programming skills, there is still a need to understand high-level concepts about technology. Unfortunately, the concepts can get extremely complex and confusing. It seems like there is an endless number of acronyms like ACL, API, OCR, CPU, HTTP, IP, JSON, NOC, PCI, RAM, and SaaS.

Even tech veterans do not know many of the terms – or have just a vague understanding of their meanings. For example, here's how Kubernetes is defined:

> *Kubernetes (K8s) is an open-source system for automating deployment, scaling, and management of containerized applications.*[1]

Huh? To get a sense of this, you really need to have a deep understanding of computer and software architecture.

But the good news is that – to use RPA effectively – there are only a handful of terms and concepts you need to know. So this is what we'll cover in this chapter.

---

[1]https://kubernetes.io/

© Tom Taulli 2020
T. Taulli, *The Robotic Process Automation Handbook*,
https://doi.org/10.1007/978-1-4842-5729-6_2

On-Premise Vs. the Cloud

The traditional IT system approach is the use of on-premise technology. This means that a company purchases and sets up its own hardware and software in its own data center. Some of the benefits include:

- A company has complete control over everything. This is particularly important for regulated industries that require high levels of security and privacy.

- With on-premise software, you may have a better ability to customize the solution to your company's unique needs and IT policies.

However, the on-premise technology model has serious issues as well. One of the biggest is the cost, which often involves large up-front capital expenses. Then there is the ongoing need for maintenance, upgrades, and monitoring. All in all, it means that the IT department may be spending valuable time on noncore activities. And finally, the use of point applications like Excel can lead to a fragmented environment, in which it becomes difficult to centralize data because there are so many files spread across the organization.

Because of all this, companies have been looking at another approach – that is, cloud computing. The interesting thing is that this has been around since the late 1950s, when computer researcher John McCarthy invented time sharing that allowed multiple users to access mainframes. This innovation would eventually result in the emergence of the Internet.

But the pervasive adoption of the PC in the 1980s and 1990s would establish the on-premise model as the preferred approach. After all, the Internet was still fairly archaic and not in wide-spread use (it was mostly for the military, universities, and large businesses). Thus, with PCs, a company could network them together to enable collaboration and sharing of data and other resources.

But as the Internet became more robust, there was a move to so-called cloud computing. One of the first business applications in this industry was developed by Salesforce.com, which made it possible for users to use the software through a browser. In the book *Behind the Cloud*, written by the company's cofounder and CEO Marc Benioff, he writes: "I saw an opportunity to deliver business software applications in a new way. My vision was to make software easier to purchase, simpler to use, and more democratic without the complexities of installation, maintenance, and constant upgrades. Rather than selling multimillion dollar CD-ROM software packages that took six to eighteen months for companies to install and required hefty investments in hardware and networking, we would sell Software-as-a-Service through a model known as cloud computing. Companies could pay per-user, per-month fees for the services they used, and those services would be delivered to them immediately via the Internet, in the cloud."[2]

The vision was bold and not easy to pull off. But as of today, Salesforce. com is the dominant cloud applications company. In fiscal 2020, the company posted revenues of more than $13 billion and the market cap was $131 billion.

But of course, no technology is perfect. So what are the downsides with cloud software? Here are just some to consider:

- With less control of the platform, there are more vulnerabilities to security and privacy lapses.

- Outages do happen and can be extremely disruptive and costly for enterprises that need reliability.

- Cloud computing is not necessarily cheap. In fact, one of the biggest complaints against Salesforce.com is the cost.

---

[2]Marc Benioff, Behind the Cloud (Jossey-Bass), 2009, p. 3.

Regardless, the fact remains that the technology continues to gain traction. According to a report from Gartner, the market for public cloud services is forecasted to jump from $214.3 billion in 2019 to $331.2 in 2022.

Here's what Gartner's vice president of research had to say: "Cloud services are definitely shaking up the industry. At Gartner, we know of no vendor or service provider today whose business model offerings and revenue growth are not influenced by the increasing adoption of cloud-first strategies in organizations. What we see now is only the beginning, though. Through 2022, Gartner projects the market size and growth of the cloud services industry at nearly three time the growth of overall IT services."[3]

Besides the impact of Salesforce.com and other cloud applications companies, another critical development was Amazon.com's AWS platform. Launched in 2005, this allowed any company to build their own cloud-native applications. The business has actually become much bigger than Salesforce.com's as revenues are expected to exceed $30 billion in 2019. To get a sense of the strategic importance of AWS, Slack is expected to spend $250 million on it for the next five years and Pinterest plans to shell out a hefty $750 million during a six-year period.[4]

AWS essentially handles the complex administrative and infrastructure requirements like storage, security, compute, database access, content delivery, developer tools, deployment, IoT (Internet of Things), and analytics (there are currently more than 165 services). This means the development of applications can be much quicker. The costs are generally lower and the fees are based on a per-use basis.

With AWS, other mega tech firms were caught off guard and scrambled to develop their own cloud platforms. The two most common ones include

---

[3]www.gartner.com/en/newsroom/press-releases/2019-04-02-gartner-forecasts-worldwide-public-cloud-revenue-to-g

[4]www.cnbc.com/2019/02/28/amazon-cloud-ceo-we-have-a-30-billion-run-rate-in-our-early-stages.html

Microsoft's Azure and Google Cloud. In fact, many companies often use two or more of these in order to provide for redundancy (this is known as a multi-cloud strategy).

The cloud also has different approaches, such as the following:

- Public Cloud: This is the model we've been covering so far in this chapter. That is, the cloud is accessed from remote servers, such as from AWS, Salesforce.com, and Microsoft. The servers have an architecture known as multitenant that allows the users to share a large IT infrastructure in a secure manner. This greatly helps to achieve economies of scale, which would not be possible if a company created its own cloud.

- Private Cloud: This is when a company owns the data center. True, there are not the benefits of the economies of scale from a public cloud. But this may not be a key consideration. Some companies might want a private cloud because of control and security.

- Hybrid Cloud: This is a blend of the public and private clouds. For example, the public cloud may handle less mission-critical functions.

As for RPA, the cloud has different implications and impacts. One is that a platform needs to deal with complex distributed applications, which can be difficult if a company develops custom programs on a cloud service.

What's more, most RPA platforms actually started as on-premise software and generally did not transition to the cloud until recently. This does seem odd as cloud computing has been around for a while and appears to be the default approach for software companies. But developing cloud-native systems is not easy for RPA as there needs to be deep hooks across many applications and environments.

In some cases, an on-premise RPA system may be loaded onto a cloud service like AWS. While there are benefits with this, it is not cloud native. This is because you will still need to upgrade and maintain the software.

So as you can see, on-premise and cloud computing are essential concepts to know. But there are some more to learn about. In the rest of the chapter, we'll cover these:

- Web technology

- Programming languages and low code

- OCR (Optical Character Recognition)

- Databases

- APIs (Application Programming Interfaces)

- AI (Artificial Intelligence)

- Cognitive automation

- Agile, Scrum, Kanban, and Waterfall

- DevOps

And then, we will take a look at how to use flowcharts, which are quite useful in an RPA implementation.

Web Technology

The mastermind of the development of the World Wide Web – which involved the use of hyperlinks to navigate web pages – was a British scientist, Tim Berners-Lee. He accomplished this in 1990. Although, it would not be until the mid-1990s, with the launch of the Netscape browser, that the Internet revolution was ignited.

At the core of this was HTML or hypertext markup language, which was a set of commands and tags to display text, show colors, and present graphics.

A key was that the system was fairly easy to learn and use, which helped to accelerate the number of web sites.

For example, many of the commands in HTML involve surrounding content with tags, such as the following:

```
<strong>This is a Title</strong>
```

Yes, this means that the text is bold.

Yet HTML would ultimately be too simple. So there emerged other systems to provide even richer experiences, such as with CSS (Cascading Style Sheets, which provides for borders, shadows, and animations) and JavaScript (this makes it possible to have sophisticated interactivity, say, with the use of forms or calculations).

No doubt, RPA must deal with such systems to work effectively. This means it will have to take actions like identify the commands and tags so as to automate tasks.

Programming Languages and Low Code

A programming language allows you to instruct a computer to take actions. The commands generally use ordinary words like IF, Do, While, and Then. But there can still be lots of complexity, especially with languages that use advanced concepts like object-oriented programming. Some of the most popular languages today include Python, Java, C++, C#, and Ruby.

To use an RPA system, you have to use some code – but it's not particularly difficult. It's actually known as low code. As the name implies, it is about using minimal manual input. For example, an RPA system has tools like drag-and-drop and visualizations to create a bot.

This is not to imply that low code does not need some training. To do low code correctly, you will need to understand certain types of workflows and approaches.

OCR (Optical Character Recognition)

A key feature for an RPA platform is OCR (Optical Character Recognition), a technology that has actually been around for decades. It has two parts: a document scanner (which could even be something like your smartphone) and software that recognizes text. In other words, with OCR, you can scan an image, PDF, or even handwritten documents – and the text will be recognized. This makes it possible to manipulate the text, such as by transferring it onto a form or updating a database.

There are definitely many challenges with effective OCR scanning, such as:

- The size of a font

- The shape of the text

- The skewness (is the text rotated or slanted?)

- Blurred or degraded text

- Background noise

- Understanding different languages

Because of all this, OCR in the early years was far from accurate. But over time, with the advances in AI algorithms, fuzzy logic, and more powerful hardware, the technology has seen great strides in accuracy rates, which can be close to 100%.

Then how does this technology help with RPA? One way is with recoding a person's actions while working on an application. The OCR can better capture the workflows by recognizing words and other visuals on the screen. So, even if there is a change of the location of these items, the RPA system can still identify them.

Something else: Automation involves large numbers of documents. Thus, OCR will greatly improve the processing. An example of this would be processing a loan. With OCR, a document will use OCR to extract information about a person's financial background, the information about the property, and any other financial details. After this, the RPA system will apply the workflows and tasks to process the loan, say, with applying various rules and sending documents to different departments and regulatory agencies.

Finally, even though RPA systems may have their own OCR, this may not necessarily be enough. Some industries and segments, such as healthcare, insurance, government and banking, still rely heavily on handwritten forms – all of which can be time-consuming and costly.

But there are OCR systems that can help out, such as HyperScience. The software leverages sophisticated machine learning (ML) technology to quickly and accurately extract the information (understanding cursive writing, for example). But there are other capabilities like detecting fields on invoices and handling data reconciliation. Consider that HyperScience can automate up to 90% of the processing.[5]

Databases

At the heart of most applications is a database, which stores data that can be searched and updated. This is usually done by putting the information in tables (i.e., rows and columns of information).

The dominant form is the relational database – developed in 1970 by IBM researcher E. F. Codd – that uses structured data. To interact with this, there is a scripting language called SQL (Structured Query Language), which was relatively easy to learn.

It was not until the late 1970s that relational databases were commercialized, led by the pioneering efforts of Oracle. Then came a smattering of start-ups to seize the opportunity. But by the late 1980s, Oracle and SAP dominated the market for the enterprise (Microsoft would essentially be the standard for the mid-market).

---

[5]www.hyperscience.com/

While relational databases proved to be quite effective, there were still some nagging issues. Perhaps the biggest was data sprawl. This describes when there is a growing number of tables that get proliferated across the organization. This often makes it extremely difficult to centralize the data, which can make it challenging to get a holistic view.

Another problem was that relational databases were not cheap. And as new technologies came on the scene, such as cloud computing and real-time mobile applications, it became more difficult to process the data.

Given all this, there emerged various alternatives to relational databases. For example, there was the data warehouse that started as an open source project in the late 1990s from Doug Cutting. The technology would undergo various iterations, resulting in the development of Hadoop. Initially, Yahoo! used this to handle the Big Data demands from its massive digital platforms. Then other major companies, like Facebook and Twitter, adopted Hadoop. The key was that a data warehouse could make it possible to get a 360 degree view of data.

The market has definitely seen lots of growth and change. Companies like Google, Amazon.com, and Microsoft have been investing heavily in data warehouse systems. There are also some fast-growing start-ups, like Snowflake, that are pushing the boundaries of innovation.

In the meantime, there have been new approaches that have gone against the model for relational databases. They include offerings like MySQL (which is now owned by Oracle) and PostgreSQL. Yet these systems did not get enough traction in the enterprise.

But there is one next-generation database technology that has done so: NoSQL. It also began as an open source project and saw tremendous growth. As of now, MongoDB has 14,200 customers across 100 countries and there have been over 70 million downloads.[6]

---

[6]https://investors.mongodb.com/

Where relational databases are highly structured, a NoSQL system is quite flexible. It's based on a document model that can handle huge amounts of data at petabyte scale.

Another major secular trend is the transition of databases to the cloud. According to research from Gartner, about 75% will be migrated.[7]

So why the cloud? For the most part, this should make it easier to allow for improved analytics and AI.

And going forward, there is likely to be much innovation with database technology. Yet relational databases will remain the majority of what companies use – which also means that this will also be what RPA interacts with as well.

APIs (Application Programming Interfaces)

An API – which is the acronym for "application programming interface" – is software that connects two applications. For example, let's say you want to create a weather app. To get access to the data, you can setup an API, which often is fairly straightforward, such as by putting together a few lines of code to make data requests (say, for the city). By doing this, you will increase the speed of the development.

APIs are pervasive in enterprise environments since they are so effective. They also have different structures. Although, the most common is a REST (representational state transfer) API.

It's true that APIs can be used as a form of automation. Yet there are some things to keep in mind:

- The technology requires having people with technical backgrounds.

---

[7]www.gartner.com/en/newsroom/press-releases/2019-07-01-gartner-says-the-future-of-the-database-market-is-the

- The development of an API can take time and require complex integration. There is also the need for ongoing testing. However, there are third-party services that can help out.

- There must be a focus on maintaining an API (it's not uncommon for an API to break if there is a change in the structure).

Even if there is an off-the-shelf API available, there are still issues. One is metering, which means that you may be limited to a certain number of requests per day or hour. Or there may be higher pricing. Next, APIs can still have bugs and glitches, especially when in complex IT environments.

Because of the difficulties, RPA has proven to be a very attractive alternative. Again, the development is much easier and there is less of a need for integration. But, interestingly enough, an RPA platform can be a vehicle for delivering advanced APIs within the enterprise.

AI (Artificial Intelligence)

A typical RPA system does not have much AI (Artificial Intelligence). The main reason is that there is a literal carrying out of tasks, which does not require any smart system. But as AI gets more powerful and accessible, RPA will increasingly start to use this powerful technology – which should greatly enhance the outcomes.

But before looking deeper at this – we will cover AI in various parts of this book – it's a good idea to get a backgrounder on the technology. Like RPA, this has also been the subject of much hype. It's not uncommon to read blogs and news articles on how AI will ultimately conquer disease, help with climate change, and even predict earthquakes! While such things could be achievable, it does seem far-fetched that they will happen any time soon. Keep in mind that AI is still in the nascent stages, even though it has been around since the 1950s!

Note   Even the much-hyped autonomous car is proving to be much more challenging than expected. Legendary Apple cofounder Steve Wozniak has noted that we'll not see this in his lifetime![8]

OK then, what is AI? Well, a good way to think about it is as follows: It's software that ingests large amounts of data that is processed with sophisticated algorithms that help answer questions, detect patterns, or learn. Interestingly enough, AI is actually made up of a variety of subcategories (Figure 2-1 shows a visual of this):

- Machine Learning: This is where a computer can learn and improve by processing data without having to be explicitly programmed. Machine learning is actually one of the oldest forms of AI and uses traditional statistical methods like k-nearest neighbor (k-NN) and the naive Bayes classifier.

- Deep Learning: While the roots of this go back to the 1960s, the technology was mostly an academic pursuit. It wasn't until about a decade ago that deep learning became a major force in AI. Some of the important factors for this included the enormous growth in data, the use of GPUs (graphics processing units) that provided for ultrafast parallel processing, and innovation in techniques like backpropagation. Deep learning is about using so-called neural networks – such as recurrent neural networks (RNNs), convolutional neural networks (CNNs), and generative adversarial networks (GANs) – to find patterns that humans often cannot detect.

---

[8]www.nbcnews.com/business/business-news/apple-co-founder-steve-wozniak-says-he-does-not-expect-n1071436

- NLP (natural language processing): This is AI that helps understand conversations. The most notable examples of this include Siri, Cortana, and Alexa. But there are also many chatbots that focus on specific uses cases (say, with providing medical advice).

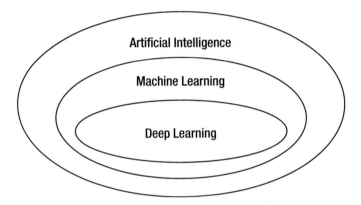

Figure 2-1. *This is a high-level look at the key components of the AI world*

When it comes to AI, the excitement is often with the whiz-bang algorithms. But the reality is that this is often a small part of building a model. Keep in mind that it is absolutely essential to have high-quality data. And this usually means spending much time cleaning it up as well as weeding out outliers, which can be a tedious manual process.

Another confusion about AI is, well, the impact from science fiction movies. These portrayals are about strong AI or AGI (artificial general intelligence), in which computers act like humans. Systems can think, engage in conversion, and walk around.

But with today's technology, it's only about weak AI. This means that the applications are focused on narrow areas, such as helping predict weather or mining insights about customers.

Besides, AI has some major issues, such as the following:

Bias: According to IBM: "Bad data can contain implicit racial, gender, or ideological biases. Many AI systems will continue to be trained using bad data, making this an ongoing problem."[9] A real-world example of this is Amazon.com, which built a recruiting system for hiring programmers. The problem? It kept selecting males! Amazon.com did change the system– and the results did not change much. The inherent problem was that much of the data – which was based on incoming resumes – were from males. In other words, AI turned out to be the wrong approach. The good news is that Amazon.com recognized this and abandoned the project. But how many cases are there where companies don't? If so, they may be engaging in unwilling discrimination. This could mean the company will miss out on attracting good candidates. Even worse, there may be legal liability.

Causation: Humans have a strong grasp of this. We know what will happen if we use a hammer to hit a glass. It's pretty much instinctive. But AI is another matter. This technology is really about finding correlations in data not causation – and this is a major limiting factor.

Common Sense: A human does not have to process many cases to understand certain rules of thumb. We just naturally understand them. But with AI, common sense has been extremely difficult to code because of the ambiguity and the lack of useful data for the seemingly infinite use cases.

---

[9]www.research.ibm.com/5-in-5/ai-and-bias/

Black Box: Deep learning can have an enormous number of layers and parameters. This means it can be nearly impossible for a person to understand why the model is generating certain results. True, this may not be a problem with facial recognition. But with applications in regulated industries, it could mean that deep learning is not viable. Consider that the deep learning systems are not allowed in financial services. Now there is more innovation in trying to find ways to understand deep learning outcomes – which is something called "explainability" – but the efforts are still in the nascent stages.

Comprehension: An AI system cannot truly understand what it is reading or observing. For example, if it read *War and Peace*, it would not be able to provide thoughts on the character development, themes, and so on.

Static: So far, deep learning has been mostly useful with constrained environments, such as with board games. There is a defined set of dimensions, objects, and rules – the kinds of things that computers work well with. It is also possible to conduct millions of simulations to learn. But of course, the real world is much more dynamic, open-ended, and chaotic.

Conceptual Thinking: AI cannot understand abstract ideas like justice, misery, or happiness. There is also a lack of imagination and creativity.

Brain: It's really a miracle of evolution. A typical brain
has 86 billion neurons and trillions of synapses. And it
only needs 50 watts a day to run! Modern computers
can come nowhere matching this power. So if AI is to
truly achieve real intelligence, there will need to be
some dramatic breakthroughs.

Then so what about AI and RPA? It's certainly a major focus right now
in the industry and we'll see many developments in the years ahead.

But for now, it's important to keep some things in mind. First of all,
there are two main types of data:

- Structured Data: This is data that is formatted (social
  security numbers, addresses, point of sale information,
  etc.) that can be stored in a relational database or
  spreadsheet.

- Unstructured Data: This is data that is unformatted
  (images, videos, voicemails, PDFs, e-mails, and audio
  files).

For the most part, RPA uses structured data. However, this represents
about 30% of what's available in a typical organization. But with AI, an
RPA system will likely be much more effective since it will be better able to
process unstructured data. For example, many companies have their own
approaches to writing invoices. Because of this, an employee would likely
have to spend much time interpreting and processing them. But AI would
be able to learn from the invoices and come up with its own rules and tasks.

Furthermore, there are other potential benefits of the technology:
judgement, the use of reasoning, and the detection of highly complex
patterns. With these, the automation will be greatly enhanced, helping
with things like detecting fraud.

Cognitive Automation

In the discussion about RPA, you may hear the term "cognitive automation" (in the first chapter, we called this IPA and was referred to as one of the flavors of RPA). It's often confused with AI – but the two concepts have different meanings.

Consider cognitive automation to be an application of AI, actually. First of all, it is mostly focused on automation of the workplace or processes in business. Next, cognitive automation uses a combination of technologies like speech recognition and NLP. By doing this, the goal is to replicate human actions as best as possible, such as by analyzing patterns of workers and then finding patterns and correlations.

Something else: Unlike other forms of AI, cognitive automation is usually effective with the use of much less data. There may also be not as much reliance on highly technical talent, such as data scientists.

Agile, Scrum, Kanban, and Waterfall

Software development can be quite complex. Besides the technical aspects, there is a need to manage a team whose members may be located in different countries. In the meantime, technologies continue to evolve. What's often even harder is maintaining a software system as there is usually a need to add capabilities and upgrade the underlying technologies.

Even some of the world's most talented software companies have faced monumental challenges. Just look at Microsoft. The company's Vista operating system took five years to develop. The reason for this was that Microsoft had too many silos within its organization and did not have sufficient collaboration. The result was that the development was glacially slow and there were persistent mishaps. The irony was that one team may have created an effective piece of code but it did not work with the complete system.[10]

---

[10]www.wsj.com/articles/SB1

In today's world, software development has become even more difficult because of the emergence of new platforms like the cloud and the hybrid cloud. This is why it's important to look at software management approaches.

One is called Agile, which was created back in the 1990s (a big part of this was the publication of the Manifesto for Agile Software Development). The focus of this was to allow for incremental and iterative development, which begins with a detailed plan. This also requires much communication across the teams and should involve people from the business side of the organization.

Nowadays, Agile has gotten easier because of the emergence of sophisticated technologies like Slack and Zoom that help with collaboration. "Over the past few years, my volume of e-mail has declined substantially," said Tim Tully, who is the chief technology officer at Splunk. "The main reason is that I mostly use Slack with my developer teams."[11]

Here are some other code development approaches:

- Scrum: This is actually a subset of Agile. But the iterations are done as quick sprints, which may last a week or two. This can help with the momentum of a project but also make a larger project more manageable (just as a side note: Scrum was first used for manufacturing but it was later found to work quite well with software development).

- Kanban: This comes from the Japanese word for visual sign or card (the roots of the system go back to Toyota's high-quality manufacturing processes). So yes, with Kanban, there is the use of visuals to help streamline the process. What's more, the general approach is similar to Agile as there is iterative development.

---

[11]From the author's interview with Splunk CTO Tim Tully on October 27, 2019.

- Waterfall: This is the traditional code development model, which goes back to the 1970s. The waterfall model is about following a structured plan that goes over each step in much detail. To help this along, there may be the use of a project management tool, say, a Gantt chart. While the waterfall approach has its advantages, it has generally fallen out of favor. Some of the reasons are as follows: It can be tough to make changes, the process can be tedious, and there is often a risk of a project being late.

DevOps

DevOps has emerged as a critical part of a company's digital transformation. The "Dev" part of the word is actually more than just about coding software. It also refers to the complete application process (such as with project management and quality assurance or QA). As for "Ops," it is another broad term, which encompasses system engineers and administrators as well as database administrators, network engineers, security experts, and operations staff.

For the most part, DevOps has come about because of some major trends in IT. One is the use of agile development approaches (this was discussed earlier in the chapter). Next is the realization that organizations need to combine technical and operational staff when introducing new technologies and innovations. And finally, DevOps has proven effective in working with cloud computing environments.

According to Atlassian, which is a leading developer of DevOps tools: "The bad news is that DevOps isn't magic, and transformations don't happen overnight. The good news is that you don't have to wait for upper management to roll out a large-scale initiative. By understanding the value of DevOps and making small, incremental changes, your team can embark on the DevOps journey right away."[12]

Note   Grand View Research predicts that the global market for DevOps will reach $12.85 billion by 2025. This would represent an 18.60% compound annual growth rate.[13]

Flowcharts

Since an essential part of RPA is understanding workflows and systems, the use of flowcharts is common. It's usually at the core of the software application.

With a flowchart, you can both sketch out the existing workflows of a department. And then from here, you can brainstorm ways of improving them. Then you can use the flowchart to design a bot for the automation.

The flowchart is relatively simple to use and it also provides a quick visual way to understand what you are dealing with. As the old saying goes, a picture is worth a thousand words.

So let's take a look at some of the basics:

> Terminator: This is a rectangle with rounded corners and is used to start and end the process, as seen in Figure 2-2.

---

[12]www.atlassian.com/devops

[13]www.grandviewresearch.com/press-release/global-development-to-operations-devops-market

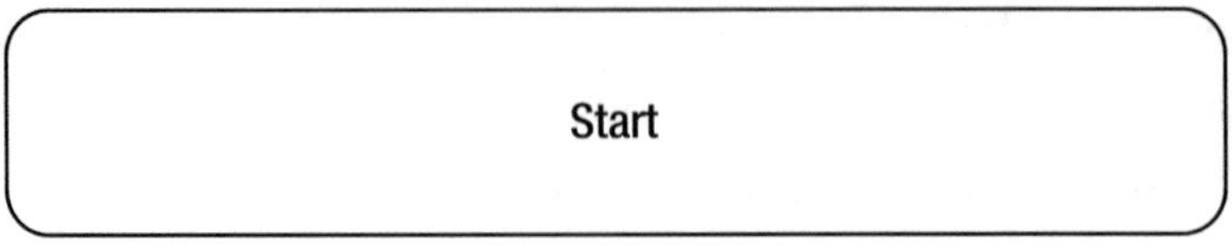

> Start

Figure 2-2. *This is a terminator, which starts and ends a flowchart*

Process: This is represented by a rectangle. With this, there is only one next step in the process. Figure 2-3 shows an example:

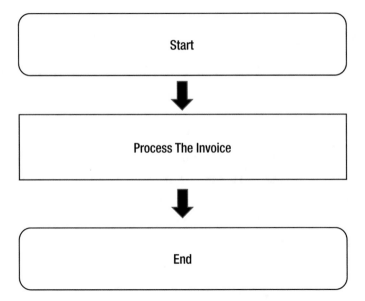

Figure 2-3. *This shows a process in a flowchart*

Decision: This is a square symbol that is at an angle. There will be at least two possible paths. Figure 2-4 is an example:

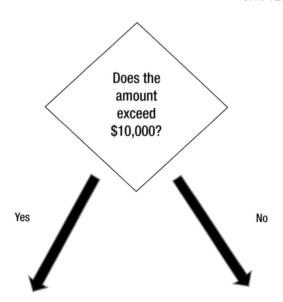

Figure 2-4. *This shows a decision process in a flowchart*

Conclusion

For the most part, you just need a general understanding of the concepts highlighted in this chapter. It's also good to remember that no technology is a silver bullet. They all have their plusses and minuses, regardless of what you might hear in the media!

So as for the next chapter, we'll take a look at the steps in implementing RPA in your organization. As you'll see, there is quite a bit of work that needs to be done before creating your first bot.

Key Takeaways

- On-premise software is where a company installs and maintains its technology within its data center. This is the traditional approach and generally allows for more control, security, and privacy. But on the other hand, on-premise software can be costly, in terms of the upfront licenses and the ongoing maintenance. The technology may also be difficult to customize.

- Cloud computing is software that is accessed via a browser. Some of the benefits include lower costs (there is no need for hardware or server purchases), less maintenance, and seamless upgrades. But the cloud has risks with security and reliability.

- There are different types of clouds. For example, a private cloud is when a company has its own data center. Then there is the hybrid cloud, which is a combination of the public and private clouds. The biggest providers of cloud services include Amazon. com, Microsoft, and Google.

- Some of the core web technologies – which allow for the creation of web pages – include HTML (Hypertext Markup Language), CSS (allows for creating borders, and animations), and JavaScript (makes it possible to have sophisticated interactivity, say, with the use of forms or calculations).

- A programming language instructs a computer to carry out certain actions. But with RPA, there is no need to learn one. Instead, a system will use low code, which involves much simpler approaches (like drag-and-drop).

- OCR (Optical Character Recognition) is software that scans and recognizes text. This technology is crucial for RPA since there is often much processing of documents.

- A database is for storing information and is an essential part of any application. The most common one is known as a relational database, which deals with structured data. But during the past ten years or so, there has emerged new types like NoSQL. They tend to work better with Big Data environments.

- An API (Application Programming Interface) is software that connects two applications. This system can provide for automation. However, it is usually tougher to develop vs. an RPA platform.

- Artificial intelligence is about processing huge amounts of data to detect patterns and find insights. The technology encompasses many categories like deep learning, Machine Learning (ML), and Natural Language Processing (NLP). As for RPA, AI is becoming an increasingly important factor.

CHAPTER 3

Process Methodologies

Lean, Six Sigma, and Lean Six Sigma

Voya, a financial services firm, pulled off an IPO in May 2013, raising $1.27 billion. This was actually a spin-off from its corporate parent, ING (the largest bank in the Netherlands). The main reason for the transaction was to help pay down the obligations from the bailout that had resulted from the financial crisis in 2009.

But with the deal, the management of Voya saw this as an opportunity to transform the organization, such as by reimagining the culture and streamlining the operations.

Note that Voya was a hodgepodge of different companies because of a long history of acquisitions. This meant that there were a variety of cultures within the organization. This made it tough to carry out comprehensive initiatives.

To help provide more centralization and a holistic culture, Voya created the Continuous Improvement Center. As a testament to its importance, the CEO of Voya led the initiative.

Keep in mind that continuous improvement is a strategy for creating better processes across an organization. And yes, this means that the effort is ongoing. There is never a true end state.

© Tom Taulli 2020
T. Taulli, *The Robotic Process Automation Handbook*,
https://doi.org/10.1007/978-1-4842-5729-6_3

But to allow for continuous improvement, there must be a focus on transparency, quality results, and efficiency (a majority of the employees at Voya have training for this). What's more, there needs to be tracking of the overall progress, such as with employee engagement, customer scores, and turnover rates. And for Voya, the ultimate goal was to find ways to better serve the customer.

All in all, the continuous improvement program has been a notable success. Some of the outcomes include the following:

- Since early 2017, the Retirement Recordkeeping Operations group has seen a 20% reduction in not-in-good-order (NIGO) applications. There were also 41% fewer escalations on hardship calls to the call center and 54% fewer were failing internal quality control checks.[1]

- The program has been the source of much innovation. For example, it helped with the creation of myOrangeMoney, which is an online interactive web site to help with understanding retirement savings. In a survey, there was an 80% satisfaction rate.

"Spending time with the culture and continuous improvement," said Jeff Machols, who is the VP of the Continuous Improvement Center for Voya Financial, "was critical in our journey with RPA. This started in early 2018 when we selected and implemented UiPath. RPA was really a natural fit with our evolution. The technology was also right for us since it did not require as much integration with our IT environment."[2]

---

[1] https://corporate.voya.com/corporate-responsibility/serving-our-clients/continuous-improvement

[2] From the author's interview with Jeff Machols, who is the VP of the Continuous Improvement Center for Voya Financial, on November 1, 2019.

But for Machols, he sees this as much more than just about technology. "Technology will not be useful unless the people understand why it's important and how it can get results," he said.

Such process initiatives like continuous improvement are common nowadays. However, they are usually referred to something like lean, Six Sigma, or lean Six Sigma. For the most part, these approaches can be quite useful when looking at RPA.

In this chapter, we'll take a high-level view of these.

Lean

In 1913, Henry Ford changed capitalism in a very big way. He introduced the moving assembly line of production, which meant that a vehicle could be made within hours not days or even months. Because of this, Ford Motor would become one of the world's most valuable companies.

But there were inherent problems with the system. It tended to be monolithic and treated workers as mere cogs. Such issues would eventually become existential problems for not only the US auto industry but many other industries during the 1970s and 1980s when foreign competition got more intense.

Of course, the main catalyst for all this was Toyota, which rethought the approach to manufacturing. In 1950, one of the family members of the company's founders, Eiji Toyoda, actually visited a Ford plant. While he was impressed with it, he realized that mass production was not viable in Japan because the country was much smaller and the customer needs were diverse. So he led an effort to come up with the so-called Toyota production system, which would allow for a seemingly magic combination of low costs, high quality, and mass customization.

But interestingly enough, in the early years, Toyota got scant attention. The company was considered a marginal player in the massive auto market and its vehicles were really substandard. But year after year, the focus on continuous improvement was leading to quality, low-cost vehicles. The irony is that US auto companies would eventually scramble to catch up and look to Toyota for guidance on how to build better cars!

For the most part, Toyota's approach was an amalgam of other process methods that would emerge. One of the most notable was Total Quality Management (TQM), which was based on the research of W. Edwards Deming. His emphasis was on the use of statistics to enhance quality. But Deming had a tough time getting the interest of US companies. Instead, he went to Japan where there was a much warmer reception.

Next, there was the development of lean. Jim Womack, an MIT Ph.D. and consultant, coined the term in the 1980s.

So then how does lean work? Why has it been so successful? Well, here's a look at some of the core principles:

- Value: For lean, value is what the customer believes is important – that is, something he or she will pay for. But this is not always obvious to determine. This is why there should be consideration of market trends and changes in tastes. There should also be a deep look at customer feedback.

- The Value Stream: Once you understand the customer value, you can then map it across development, production, and distribution. Along this process, the focus is on finding ways to eliminate the waste. Essentially, these are any factors that reduce value, such as long wait times, quality issues, and high transportation costs. Yet there is some waste that is necessary, such as in the case of meeting regulatory requirements or ensuring the safety of a dangerous process.

- Flow: Even if a product or service has value to the customer and has minimal waste, there still needs to be ways to make sure there is efficient creation and delivery. To do this, you can break down the process into small steps and find ways to optimize them. This also needs to be pervasive across the organization. If a process is only for one department, then its usefulness may be negligible.

- Pull: Inventory can often be the biggest source of waste. It's expensive and can be a drain on the attention of the organization. With pull, the approach is to produce quantities when needed, in a just-in-time framework. And for this to work, there needs to be a strong understanding of the customer value.

- Perfection: This is considered the most important step. This involves the constant pursuit of continuous improvement, which should be the goal of every employee (this comes from the Japanese word kaizen). Employees must be empowered to take actions to pursue continuous improvement.

To get started with lean, there needs to be an understanding that it requires a change in mindset. It's really about looking to the long term, not quick fixes. The reason is that there needs to be a true change in the culture of the company.

But as with anything new, a good approach is to take a first look at one problem to solve. This will allow for a learning experience. Then as the team gets more familiar with lean, there can be more ambitious projects to take on.

Moreover, communication is absolutely critical. To this end, you can use a software platform like Slack, Asana, or SmartSheet, which provides collaboration and tracking. They also do not require the training for traditional project management.

"It's simple but there must be clarity of objectives and plans," said Dave King, who is the Chief Marketing Officer at Asana.[3] "But such things can get lost in an organization. It's common for employees to not know the main goals. But with a product like Asana, everyone can be on the same page, which means a company is more agile. And yes, we have seen great results within our own organization. For example, we have several hundred community events every year and each requires 115 tasks. Without a template to track and automate the process, we would not be able to have so many events."

Lean can be kind of fuzzy, with many variations and methods. It's actually common for a company to create its own version, which fits its unique needs.

A famous illustration of this is Danaher. The company got its start in 1969 as a real estate investment trust. But its founders, Mitchell Rales and Steven Rales, would eventually transition out of this business and use the company as a vehicle for acquisitions.

It began with the merger of Jacobs Manufacturing with Chicago Pneumatic, to form a company that manufactured brakes for diesel trucks and power tools and industrial equipment. However, the company was under much pressure and was even headed for bankruptcy. The Rales brothers realized they had to do something dramatic – and very quick. So they sought out the help of two Toyota production experts, who inspected the factory. As should be no surprise, the Rales brothers got an earful. The bottom line: the factory was an absolute mess.

---

[3]This is from the author's interview with the CMO of Asana, Dave King.

But the Toyota experts knew it could be saved, so long as there was a complete rethinking of the processes. For example, one change was to go from a clockwork to counterclockwork assembly. Keep in mind that – on average – a person's right arm is 3 percent stronger, which translates into higher productivity! There were also a myriad of other lean processes to provide for continuous improvement.

All in all, the results were standout. Danaher would then use the learnings to create its own methodology, called the Danaher Business System (DBS). The company summed it up as "DBS engine drives the company through a never-ending cycle of change and improvement: exceptional PEOPLE develop outstanding PLANS and execute them using world-class tools to construct sustainable PROCESSES, resulting in superior PERFORMANCE. Superior performance and high expectations attract exceptional people, who continue the cycle. Guiding all efforts is a simple philosophy rooted in four customer-facing priorities: Quality, Delivery, Cost, and Innovation."[4]

To support this, Danaher established kaizen sessions and policy deployment reviews (these are extensive meetings to review the progress on goals) across all the divisions. Every day, the question to ask was: "How can things be made better?"

As a testament to the priority of DBS, the CEO and executives would teach courses about lean for two weeks out of each year. In fact, the former CEO, Larry Culp, wrote daily commentary on the company intranet about the kaizens. He would also lead an annual conference to go over the best practices.[5]

So what has been the impact? It's been stunning. Since the first acquisition in 1986, Danaher's market value went from $400 million to $97 billion (this does not include the $23 billion value of the other part of the business, Fortive, that was spun off in 2016).

---

[4]www.danaher.com/how-we-work/danaher-business-system
[5]www.strategy-business.com/article/Danahers-Instruments-of-Change?gko=12606

Six Sigma

Some of the core concepts of Six Sigma go back about 100 years, with the usage of statistical approaches for manufacturing. But it was during the mid-1980s that Motorola set out to better formalize the methods and also add its own. As the company showed notable improvements (the initial focus was on its pocket pager business), the use of Six Sigma began to gain much popularity. However, it was GE CEO Jack Welch who became the most high-profile advocate. He once noted: "Six Sigma is a quality program that, when all is said and done, improves your customer's experience, lowers your costs, and builds better leaders."[6]

For the most part, Six Sigma was about looking at disciplined ways to greatly improve quality by detecting defects, understanding their causes, and improving the processes. All this would have to be repeatable and sustainable – and yes, based on data.

The term Six Sigma comes from statistics, which represents six standard deviations from the mean. Figure 3-1 shows how it looks visually:

---

[6]http://asq.org/quality-progress/2016/10/standards-outlook/a-giving-tree.html

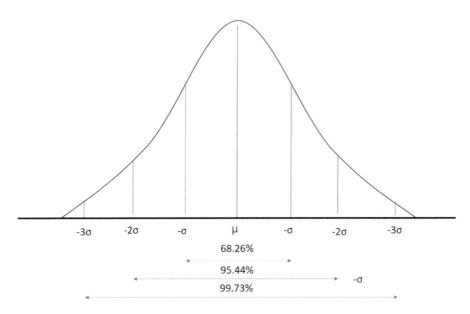

Figure 3-1. *This is a chart of Six Sigma, which shows six standard deviations from the mean*

To understand this, let's get a short backgrounder on statistics. Figure 3-1 is a normal curve or a bell curve. It represents the sum of probabilities of data in a sample population and the midpoint represents the average or mean.

In the natural world, the normal curve actually reflects much of what happens, such as with height and weight. Thus, if the data is within the first standard deviation, then the data will fall within 68.26% of the total, and so on.

When it comes to Six Sigma, the normal curve means that if a process is within six standard deviations, then the process has minor defects. This is at 3.4 defects per million opportunities (DPOM). In other words, it is virtually nothing. But of course, it is extremely tough to attain this level. For many processes, it is 3 to 4 sigmas. By using techniques of Six Sigma, the focus is on finding ways to improve on the DPOM.

When doing this, there needs to be a way of measuring different processes, to get a sense of the progress. One way of doing this is using defects per opportunity (DPO), which is calculated as follows:

Total number of defects/total number of defect opportunities

Let's take an example. Suppose we are looking at two processes in a car company. One is the attachment of mirrors, which involves 20 steps. Then there is a much more complex process – for building an engine – that has 1,000 steps.

Suppose a report shows that in a recent production run, there were 5,000 defects for each of the processes. But of course, this is really a false comparison since the two processes have much different levels of complexity.

To deal with this, we can use the DPO calculation.

Mirrors:

$5,000 / (20 \times 5,000) = 0.05$

This means there are 50,000 defects per one million.

Engines:

$5,000 / (1,000 \times 5,000) = 0.001$

In this case, there is a much better result, with a mere 1,000 defects per million.

Table 3-1 shows how this is done. You can see that it is not a linear improvement – but instead exponential. For example, going from 3 sigma to 4 sigma results in a 10X gain!

Table 3-1. *Six Sigma Levels*

Sigma Level	DPMO	% Good
1	690,000	31.000%
2	308,537	69.1463%
3	66,807	93.3193%
4	6,210	99.379%
5	233	99.9767%
6	3.4	99.99966%

Table 3-1. *This shows how Six Sigma can improve on effectiveness.*

While the roots of Six Sigma were with manufacturing companies, the method has proven quite adept. It has been used for a wide variety of other types of companies, both large and small. The reason is that there are always opportunities to improve processes to increase effectiveness and efficiency.

How to Implement Six Sigma

There are different approaches to apply Six Sigma to a process. But one of the main ones is DMAIC, which stands for define, measure, analyze, improve, and control (it is pronounced as duh-may-ick).

Here's a look:

- Define: You will assemble a team and come up with a name for the project. The next step will be to identify a problem to solve (such as increasing delivery time or improving customer satisfaction scores). There should also be a timeline and clear-cut milestones to achieve. To help with this, it's a good idea to come up with a written plan, which is often referred to as a project charter.

- Measure: A good place to start is to map out the process, which can be done using a flowchart. It's also advisable to get others to review it. After this, you can put together a data plan, which shows what data to collect and how to obtain it. Of course, this process can be time-consuming – but it is important to make sure it is done right. If the data is off, so will your results.

- Analyze: This is more of a subjective part of the process. That is, you want to get a sense of the root causes of the problem. What is really going on? Why is the problem really happening? You can have brainstorming sessions as well as use diagraming systems, like the fishbone diagram. Statistical techniques can also be helpful like regression (this is a way to determine the correlation with certain variables).

- Improve: At this stage, you will devise solutions to the problem. But you need to be cautious as the solution may have land mines, such as unintended consequences. So there needs to be a look at error-proofing the solution and using quality control processes. A helpful method is FMEA (failure modes and effects analysis), which was initially used in the US military during World War II and was also crucial for the Apollo space program. This is about looking at all steps and trying to identify the potential failure points.

- Control: Once you have the solutions, you want to make sure they remain in place. This means establishing monitoring systems as well as having ongoing reviews to find better ways to improve the processes. There is also a need to understand when change is truly needed. After all, processes have random variances that do not require action. Thus, one way to help identify where there should be change is to use a statistical process control (SPC) chart. This graph has three lines: the average, the upper line, and the lower line. These are charted over time, which should allow for a visual way to get a sense when the variances are notable and require action.

While Six Sigma can be powerful, it is certainly not for everything. There is simply not enough time and resources to do so. In other words, it is critically important to look at those processes that will have outsized impacts on the organization. At the core of this is to think about the customer – or, as is known in Six Sigma, the voice of the customer or VOC. At first, you need to think of things in the language of the customer. How is he or she viewing the problem? After this, you will translate it for your organization into something called CTQ (critical-to-quality) requirements. These must not only be very important but also measurable, such as in wait times, quality of the product, and on-time delivery.

Yet lean as a general philosophy and strategy has still proven to be quite effective across many different types of use cases. For example, in a study from the American College of Emergency Physicians, it was found that emergency rooms that implemented lean thinking showed improved patient care quality, wait times, and the number of patients who left the waiting room.[7]

---

[7]www.ncbi.nlm.nih.gov/pubmed/21035904

Six Sigma Roles and Levels

Six Sigma has various roles for the project. They include the following:

- Executive: This essentially shows that there is significant buy-in, which should lead to traction. The executive will set the overall objectives and provide approvals for key actions.

- Champion: The executive will select this person, who will have Six Sigma training. The champion will take on the operational management of the project, such as by providing the necessary resources and removing any roadblocks. There will also likely be a need to manage across different departments.

- Process Owner: Each of the key processes should have such a person. He or she will help manage the team and make sure the project is on track. To do this, a process owner needs to have experience with statistical techniques.

Besides these roles, there are different levels of proficiency for Six Sigma, which are based on the concepts of martial arts. Here's what they are:

- White Belts: This is a novice at Six Sigma and will have taken a few hours of training, such as on the fundamentals of quality and process thinking.

- Yellow Belts: This also means having a basic level knowledge of Six Sigma, covering areas like the DMAIC system, variation, and process mapping. A yellow belt may be an executive or champion.

- Green Belts: This is a person who works with black belts and will help put in place Six Sigma systems. This will require training on statistics and data analysis (which can take a few months).

- Black Belts: This is a full-time Six Sigma specialist who has an understanding of advanced statistics.

- Master Black Belts: As the name implies, this is the highest level for Six Sigma. They will not only help with managing a project but also provide training and mentoring.

Consider that there is no organization that provides an official Six Sigma certification. Instead, there are a plethora of training options available (as a quick Google search will show!), spanning from online to classroom instruction. There are also plenty of good books on the subject, such as *The Six Sigma Handbook* and the *Lean Six Sigma Pocket Toolbook*. Although, the Council for Six Sigma Certification does provide for standards for training.

Despite all this, to have a successful Six Sigma implementation, you do not necessarily need the different belts. Instead, the key to success is making sure the team understands the main principles of the approach.

Finally, there are software packages that can help, such as Minitab. With this, you can create charts as well as engage in hypothesis testing, compute regressions, and other statistics.

Lean Six Sigma

Because lean and Six Sigma are so powerful – and have similarities – there is something called lean Six Sigma! This is another process methodology that takes some of the best approaches of each. That is, there is the statistics and data techniques of Six Sigma as well as the focus on lean's elimination of waste. Lean Six Sigma came about in 2001, with the publication of the book *Leaning into Six Sigma: The Path to Integration of Lean Enterprise and Six Sigma* by Barbara Wheat, Chuck Mills, and Mike Carnell.

So then, how to go about using lean Six Sigma? A common framework is to first start a lean implementation. For example, you can use the 5 S's, which include the following:

- Sort (Seiri in Japanese): Get clarity by taking away needless processes, clutter, and items.

- Straighten (Seiton): This is about using storage and resources efficiently, such as by picking the right item at the right time. To help with this, you should organize everything in a way that makes it easy to take actions. This could actually be something like setting up an office space for maximum efficiency (not just the factory floor).

- Shine (Seiso): Simply put, make sure everything is clean and tidy, which must be a daily activity. There should also be efforts to identify the root causes of dirtiness.

- Standardize (Seiketsu): Come up with a step-by-step process for a clean and neat workplace, which should be backed with clear roles and responsibilities.

- Sustain (Shitsuke): With the standards established, there must be ways to make sure they are upheld and maintained. This can be tough as an organization can easily lose interest in the methods.

Along the way, you want to always find ways to reduce and eliminate wastes. In lean, there are actually seven different types of wastes to consider (they were developed by Taiichi Ohno, who created the Toyota Production System):

- Motion: As we saw with Danaher earlier in this chapter, the organization of a workflow can result in much wasted movement – which will lead to higher costs and more delays. When it comes to motion, there needs to be a detailed look at both the people and machines.

- Transportation: This is often a major source of waste. Interestingly enough, start-ups like Lyft and Uber are based in part on dealing with this. And yes, the trucking industry has adopted different techniques and technologies to bring more efficiency. And just look at what Amazon.com has been doing with its own delivery infrastructure, which is likely to rival those of FedEx and UPS.

- Defects: There are different types. A design defect is where a product has a problem because of ineffective development or testing. Then there is a manufacturing defect, which is when the problem is the result of a flaw in the assembly. And then there are defects about the elements of a product (an example is the use of asbestos in buildings that resulted in substantial liabilities). But regardless of which one, they all are something that need to be guarded against. Even a seemingly small defect can cause much trouble for a company. This is why it's important to have strong processes in place to catch defects.

- Overproduction: This waste can easily wreck a company. The toy industry is particularly vulnerable to overproduction as are apparel companies. Let's face it, consumer tastes can change on a dime. But there are definitely ways to help with the risks. By using sophisticated software and analytics, it's possible to get a better sense of how consumer demand is evolving. This can also be combined with just-in-time manufacturing.

- Inventory: This is similar to overproduction but also includes supplies and work in progress. If too much is purchased or created, the heavier the costs. There is also the potential that the inventory will ultimately have to be written off. A company that showed how the effective management of this could turn into a superior business advantage was Dell Computer during the 1990s, which innovated the build-to-order strategy. The company would take orders directly – through the phone and eventually from the Internet – and then the PC would be quickly assembled. The suppliers would also be paid later, which would generate higher cash flows. The result was that Dell did not suffer from the heavy expenses of inventory accumulation and was able to fend off many competitors to become the top producer in the industry.

- Waiting: This waste can add up quickly, whether in terms of connections with suppliers, partners, or customers. In today's world, the expectation is that service should be as quick as possible. It's something that has been driven by companies like Amazon.com.

- Overprocessing: This is where any part of the process of the manufacturing or development is unnecessary. This could be something like creating an elegant design for an engine, even though most customers would not notice or care about them.

Once you have put in place mechanisms for lean and they begin to get results, you can then look at Six Sigma. You could start with something like kaizen or DMAIC, which we have covered earlier in this chapter.

But there are certainly a myriad of others to look at. For example, there is something known as poka yoke (this is from the Japanese words for avoid and mistakes). The goal is to create a process that has minimum errors. This could mean putting together a checklist for employees or some type of automation with a computer app, such as RPA.

Finding the Right Balance

As with anything, process methodologies can be taken to the extreme, which could lead to awful results. This is usually the case when management is mostly focused on driving steep cost cuts. True, this may get short-term results, but the moves could hamper a company's ability to compete.

Consider the case with Kraft Heinz, which was the result of a transformative $49 billion merger that was struck in March 2015. The backing came from some of the world's top investors, including 3G Capital Partners and Warren Buffett.

Actually, this is what Buffett had to say about the deal: "I am delighted to play a part in bringing these two winning companies and their iconic brands together. This is my kind of transaction, uniting two world-class organizations and delivering shareholder value. I'm excited by the opportunities for what this new combined organization will achieve."[8]

---

[8]http://ir.kraftheinzcompany.com/news-releases/news-release-details/ hj-heinz-company-and-kraft-foods-group-sign-definitive-merger-0

The managers at Kraft Heinz emphasized ways to cut costs and streamline operations. And for some time, the actions worked.

But they ran too deep. The reality was that Kraft Heinz scrimped too much on R&D, marketing, and product development. Essentially, the company was being squeezed – and hard. Even worse, the consumer products market was undergoing major changes, as customers were looking for healthier offerings and lower priced alternatives to premium brands.

By February 2019, Kraft Heinz disclosed the $15.4 billion write-down of the Kraft and Oscar Mayer brands. There was also a slashing of the dividend. Since the merger, the stock price has plunged from $80 to $32.

According to business management author and professor, John Kotter, in a Harvard Business Review article: "Ultimately, leaders need to understand that the pace of change is accelerating everywhere, not just in packaged foods. In our view, Kraft Heinz's experience shows dramatically that traditional methods of restructuring are increasingly risky. Any effort that slows down or curtails a company's ability to innovate can lead to disastrous results."[9]

Applying Lean and Six Sigma to RPA

The uses of lean, Six Sigma, and lean Six Sigma demand a major commitment of time and resources. It could easily mean delaying an RPA implementation for six months to a year. But this may ultimately be worth it. Keep in mind that many of the top IT and management consulting firms will have many Six Sigma black belts who can help redesign a company's processes, which should go a long way in making an RPA system even more impactful.

This is not to say you need to retain such a firm (although, in the next chapter, we will cover the factors to think about this approach). There are many examples of companies that have taken a do-it-yourself approach and have been successful with it.

Regardless, there still needs to be some level of attention to current processes and standard operating procedures. The fact is that they were likely created in an ad hoc fashion, with not a lot of thought! Because of this, there should be rich opportunities to find enhancements.

But there are definitely some takeaways from lean, Six Sigma, and lean Six Sigma that can help with these preliminary actions. First of all, it's advisable to put together some process maps, such as with a tool like Microsoft's Visio or IBM's Blueworks Live. The software will allow for quickly creating visualizations that will highlight the potential areas for redesign. You will also be able to invite other users to get valuable feedback.

When doing this, there are some factors to keep in mind:

- What are the bottlenecks and root causes? What can be done to improve them?

- Think about the core ideas of FMEA. So before making a change to the process, you need to brainstorm about the unintended consequences. Where could things go off the rails? Ultimately, you might realize that some functions are just poor candidates for automation.

- Motion: Do not just use the software. Walk around the office and get a sense of the organization. Then go through the physical aspects of what an employee does at the desk. By doing all this, you should get a much better idea of where the opportunities are to improve the processes.

- Implications of Automation: As people are prone to making mistakes – especially with tedious tasks – there may be safeguards put in place. This may be something like having someone review a process. However, with RPA, this step is probably not needed as the automation will carry out the process the same every time. But this is so long as there is extensive testing, validating, control, and audit of the bot before it is released into a production environment.

- Authority: In a process, there are often times when there needs to be approval from a manager. This could be for allowing the issuance of a payment, say, over a certain amount. But with RPA, this process can be streamlined by use of exceptions that are built in to the system. What's more, with emerging AI technologies, it may even be possible to create a bot that can make the decision on its own.

Once you have your process map, you will be in a much better position for your RPA implementation. But in keeping with the principles of lean and Six Sigma, it's also important to have a focus on continuous improvement. The process map should just be the first step in a journey to keep finding more ways to improve processes.

Oh, and yes, there should also be an emphasis on: What is the best for the customer? How will the change in the process improve value?

Conclusion

Again, as noted in this chapter, you do not have to be a lean or Six Sigma expert for improving processes within your organization. But you can use some of the techniques to make improvements, which should allow for a better RPA implementation.

As for the next section, we'll take an in-depth look at how to implement a system, which will involve planning, bot design, deployment, and monitoring.

Key Takeaways

- Lean is a process methodology whose roots go back to Toyota in the 1950s. The company needed to find a unique way to produce cars that combined low costs, high quality, and mass customization. Over the years, the company experimented with different approaches, which eventually focused on continuous improvement.

- It was during the 1980s that lean was coined and formalized. Some of the core principles included value (what customers will pay for), value stream (the mapping of the development, production, and distribution), flow (efficient delivery of value), pull (production that is only needed), and perfection (the pursuit of continuous improvement, which should be the goal of all employees).

- For lean to have true impact, there must be a change in mindset. It's about taking the long view of things, not just looking for quick fixes.

- Six Sigma also goes back many years. But it was during the mid-1990s that it became a well-defined framework. This was because of the efforts of Motorola, which used Six Sigma to greatly improve its manufacturing processes.

- Six Sigma is generally about using statistics to minimize defects. The goal is to reach near perfection – or 3.4 defects per million opportunities.

- A common way to implement Six Sigma is to use a framework called DMAIC. It involves the following: Define (look at the problem to be solved), measure (map the process and create a data plan to track things), analyze (look for the root causes of the problems), improve (look for more efficient solutions but also try to make sure there are no unintended consequences), and control (this is about making sure the solutions remain in place).

- In Six Sigma, the Voice of the Customer (VOC) is where you think of a problem from the perspective of the customer. This is then translated to the organization through something called CTQ (Critical-to-Quality) requirements.

- Six Sigma has a various roles: Executive (this is a senior-level person who sponsors the project and sets the objectives), champion (this person has Six Sigma training and will provide operational management), and process owner (this is a person who will use more advanced Six Sigma techniques to manage a team).

- Six Sigma has different levels of proficiencies described in terms of martial arts – from white belts to master black belts.

- Lean Six Sigma combines the benefits of both process methodologies. With lean, you get the focus on the elimination of waste and other inefficiencies and Six Sigma helps with data and statistics. A typical approach is to first use lean and then go to Six Sigma.

CHAPTER 4

Planning

Building a Solid Foundation for Your RPA Implementation

Ernst & Young (EY) is a strong believer in the power of RPA. The firm not only has a thriving practice in the industry but also uses the technology to improve its own operations, on a global scale.

But this does not mean that EY is not without its criticisms of RPA. The firm realizes that – as with any type of major enterprise software system – there are many potential land mines. Of course, EY's consultants are good at navigating clients through them. Interestingly enough, one of the firm's fastest-growing businesses is providing remediation services for failed RPA implementations!

In a report from the firm, there was an extensive analysis of a broad-range of RPA implementations (spanning 20 countries). The high-level conclusion? Well, it was not encouraging. EY found that anywhere from 30% to 50% of initial RPA projects failed.[1]

Now this is not an indictment of the technology. Note that EY believes that RPA vendors do generally have solid software. But as with any technology that has a potential major impact on an organization, there needs to be considerable planning. But unfortunately, this is something that can get scant attention.

In this chapter, we will highlight the initial steps of RPA, so as to put together a plan that will provide for a durable foundation.

---

[1] www.ey.com/Publication/vwLUAssets/Get_ready_for_robots/%24FILE/ ey-get-ready-for-robots.pdf

The Preliminaries

It's true that RPA is relatively easy to use because of the interactive workflows and drag-and-drop capabilities. But this can lead to a nagging problem – that is, it is tempting to just rush and get started with an RPA implementation, such as by creating many bots to automate processes.

On the surface, this approach has some intuitive advantages. Hey, after all, shouldn't an organization be agile? Isn't it a good idea to experiment and try innovative ideas?

These are all important. But if you are too eager to implement RPA, the outcome can easily be chaos and failure.

Rather, a preferred approach is to draft a written plan of action, which provides the key priorities, objectives, and roles for the RPA implementation. There should also be a look at security, the impact on IT, governance, and compliance.

As should be no surprise, a good exercise to start with is brainstorming. At this stage, be freewheeling and do not criticize ideas. The goal is to unleash creativity and ingenuity. As much as possible, try to get feedback from different parts of the organization, such as marketing, finance, HR, legal, IT, and sales.

Some of the questions to consider include the following:

- What is the degree of automation in your department?

- What works already? Why?

- What is falling short?

- What processes can be improved? And can automation help?

- What are the processes that are repetitive and routine? How many people are involved? The time spent on the processes?

- What technologies have not worked in the past? Why?

- What is the general sentiment about automation? Is there resistance to this? Might employees fear for their jobs?

You will essentially be in a discovery mode. So make sure you have an app that can track the ideas and allow for commenting. Something as simple as putting together a Google doc can help with the process.

For some companies, they will set up a workshop, say, for a few days. It will be an intensive focus on ideation. But it is also an opportunity to provide some training, which would include tutorials on the fundamentals of RPA, the benefits of the technology, and how it will change daily work.

To get a real-world example of a workshop, consider CloudStorm. The company is a provider of services to help RPA software but also sets up one-day workshops for its clients.[2] Here's the structure:

- Introduction to software bots and automation: This is a general primer on RPA and how it is used. There is also coverage of different use cases that are relevant to the client's industry. Finally, CloudStorm goes over what tasks are the best for automation as well as some of the gotchas to avoid.

- Process Automation Identification and Elaboration: This involves group activities to come up with ideas to identify areas to automate. By doing this, you can have a white board or wall where you pin them. After some discussion, the team can then score the ideas (in terms of feasibility, cost, and potential return), such as by voting.

- Discussion and Validation: This is another interactive discussion. You can divide the group into different teams and then present and validate the ideas.

---

[2]https://cloudstorm.io/process-automation-workshop/

The next step? CloudStorm will draft a report with the results of the workshop and provide next steps: Suggest areas to automate, estimate the expected impact, and craft the KPIs.

Finally, some companies will use software to help with the mapping and optimization of processes. This is something known as process mining. We'll cover this in detail in Chapter 12.

Use a Consulting Firm?

It's common for a company to use a consulting firm when planning for an RPA implementation. Granted, this may wind up costing more and yes, it can be tough to find a qualified consulting firm.

Or in some cases, the consulting firm may not have the right team to help you out, such as by having direct experience with your industry or company size.

Despite all this, there are certainly some clear advantages of hiring a consulting firm. In fact, the costs should be well worth it if you have the right partner. It's also important to keep in mind that a consulting engagement does not have to be long term. For example, in the case of Symantec's RPA implementation, the company retained a consulting firm for the first couple months so as to help with the initial planning and bot development. After this, the company's own internal team took over the project.

In terms of the benefits of using a consulting firm, here are some of the main ones:

- The company should have experience in identifying the right areas to automate. For example, as the firm will have seen the full life cycle of an implementation, it will understand what to anticipate and the danger points. This can be a major factor in a successful RPA implementation.

- You and your own team are likely too close to the company's processes. There may actually be some glaring opportunities that could be missed! Or, in some cases, there could be areas for automation that may be far from obvious. Yet a good consultant will have a deep understanding of the technology so as to know what can realistically be accomplished.

- A consulting firm can prove invaluable in the evaluation and selection process for an RPA platform and other related software. But of course, you need to be cautious. A firm may specialize in one platform or even have some type of financial arrangement with the partner, which means you may not be getting the best option. So when interviewing a consulting firm, see if they are vendor agnostic.

- Some consulting firms – such as the larger ones – will have their own proprietary frameworks and methodologies. This involves a standard set of strategies that have worked in other industries and that should be more understandable to people who will use the RPA software.

- A consulting firm may have created its own software to help with the RPA implementation. Some examples of this include systems to evaluate your company's processes, prebuilt libraries, and bots.

- A consulting firm can be helpful in looking at taking your RPA implementation to the next level. This would be the case with the use of AI, which usually takes much time and resources for an organization to adopt. But a consulting firm can help smooth out the process and get to actionable results faster.

- A consulting firm can be a good source of training and should also have educational materials for self-learning. It's also a plus if there are interactive tools and videos.

- A Center of Excellence (CoE), which we'll cover in Chapter 6, is a team that manages the RPA system. And yes, a consulting firm can help assemble this team and may even be a member of it.

- A consulting team should have in-house capabilities to handle the complex issues of compliance and regulations, which can be included within the RPA design. There can also be help with setting up the right structure for governance (risk assessment, oversight, and controls).

- The business model may be based on results. That is, there will only be payments if certain milestones are reached, allowing for more alignment of goals.

- A consulting firm may provide a "robotics-as-a-service" option. This means that a client can host their RPA system on the firm's servers, which will likely lower the costs and make it easier for managing the project.

True, there are many RPA consulting firms. This is inevitable as it is not necessarily difficult to start one and the industry is growing at a rapid pace. This definitely adds to the risks. In other words, make sure you take the time to interview a variety of firms and check the references.

It's also critical to have a firm that is either local or willing to travel to your location. Having a virtual-based RPA consulting arrangment is probably not good enough. A major reason for this is that it will be tough to evaluate your company's processes.

> **Note**   In Appendix A, there is a list of the top consulting firms that have specialties with RPA.

RPA Consulting: Some Case Studies

For a large organization, an RPA implementation can be a major undertaking. But a consulting firm can definitely be a big help, greatly increasing the odds of success.

Let's take a look at a case study of Cognizant, which helped a major insurance company with its RPA initiative. Like many others in the industry, the company had a large number of manual processes that could be automated.

At first, the focus was on the operations in Australia with the HR and finance departments. Yet it was a significant endeavor as Cognizant was tasked with assessing the back-office processes of about 2,000 employees. There were extensive conversations with leaders in the departments that helped to narrow the number of areas to target for automation. By doing this, Cognizant identified more than 100 automation opportunities – with the potential for $7 million in net savings.[3] This meant that more than 80 people could be redeployed for more value-added work.

However, in order to make the project manageable, Cognizant narrowed the list and then conducted an in-depth analysis of the remaining ones. This also resulted in a detailed report, which set forth the goals, the deliverables, and ROI expectations.

This is not to imply that the insurance company could not have done this on its own. Although, it seems reasonable that the process would have been longer and there would have been more mistakes made.

---

[3] www.cognizant.com/case-studies/pdfs/robotic-process-automation-for-insurance-company-codex3250.pdf

For other case studies of consulting success, consider the following:

- Genpact, which is a top consulting firm, helped a large
  aerospace firm with its RPA implementation (it also
  included the addition of AI). Some advisory services
  included help with assessing the company's processes,
  selecting the software, designing the CoE, and putting
  together a pilot program. The results? The aerospace
  company recorded productivity savings of 30% to 50%
  and the expectation is that there will be an 8X return on
  investment (ROI) over a three-year period.[4]

- A global bank wanted to find ways to manage higher
  volumes of transactions without hiring more employees.
  Keep in mind that the processes were generally complex
  and the data was scattered across silos. To help with all
  this, the bank retained Deloitte to provide an analysis of
  an RPA project. In only six weeks, the first bot was created
  and deployed successfully. Realizing the quick benefits,
  the bank aggressively expanded on the effort, with bots for
  credit card remediation, PDF conversion, and payment
  processing. The bottom line? The bank had more than
  150 bots that processed 120,000 requests per week at
  only 30% of the costs of using employees. The estimated
  savings were about $40 million for the first three years.[5]

- A global rental car company wanted to automate its
  shared services center in Budapest. For the most part, the
  use of Excel and basic scripting was no longer effective.

---

[4]www.genpact.com/insight/case-study/one-small-step-for-an-aerospace-firm-is-a-giant-leap-forward-for-intelligent-automation

[5]https://www2.deloitte.com/content/dam/Deloitte/us/Documents/process-and-operations/us-cons-rpa-global-bank-case-study-infographic.pdf

The firm retained EY for an RPA pilot program, helping with identifying areas to automate, selecting the right tools, and building bots. The potential savings amounted to 10 to 20 full-time equivalent (FTE) employees or €1 million over three years – for a 900% ROI. Because of the success, the car rental firm has looked at extending the project for 200 to 400 FTEs over two years.[6]

What to Automate?

It seems easy to select what needs to be automated, right? Well, this is true in some cases. But as we've seen already in this chapter, there needs to be some analysis of existing processes – to get a holistic view of the workflow. From this, you will be in a much better position to see what needs to be automated.

But it is also critical to understand that RPA is a fit for certain types of automation. It's not an one-size-fits-all solution.

What's more, when evaluating what to automate, there should be a group of people with diverse backgrounds – in terms of technology, business expertise and knowledge of the day-to-day operations of a department.

Then what are the areas that are right for RPA? Here's a look at some key factors:

- Tedious work: It's the kind of activity that requires little knowledge. This could be cut-and-paste, and clicking a variety of buttons. Let's face it, do you want to pay a talented employee do such mundane things?

---

[6]https://www.ey.com/uk/en/services/advisory/ey-robotic-process-automation

- Time-consuming: Which types of processes result in late outcomes? This could be, say, for creating a report or doing end-of-quarter processing.

- Repetitive: The process has a set of steps that rarely change. This is ideal for the mechanical approach of RPA.

- Frequency: To get real value, the process is something that is recurring. This may be where a person engages in the activity a few times a week.

- Rules-Based: This is because RPA allows you to easily create workflows, which have features like IF/Then logic to carryout actions on a consistent basis.

- Clearly Defined Processes: You want to make sure you have some basic metrics, such as the number of steps or a flow chart.

- High-Volume: Such activities can be problems for workers. But of course, software is spot-on for handling high-speed processes.

- Prone to Error: Are there areas where data input errors are common?

- API: If an API is not available, then the process could be a good candidate for RPA automation.

- Customization: True, your apps may allow for this. But it is costly. You may have to pay fees for the vendor as well as for hiring third-parties. In other words, the RPA option could be more cost-effective.

- Sensitive Data: Access to employees is certainly a risk. But with part of a well-defined RPA bot, you should have much better protection of the data.

- Scale: If you do not want to hire more people to ramp up operations – because the current IT setup is fairly rigid – then RPA could be a good option.

- Organization: Look for areas within a company where there are persistent silos and bottlenecks.

Yes, this is a lot! And it can be overwhelming. This is why you might want to narrow the criteria down to four to five factors. For example, the common ones used include: tedious work, rules-based, high-volume and repetitive.

Next, for each of the processes you are considering, you can put together a diagnostic. Example: Suppose you are looking at automating the customer cancellation process. You might have the following:

Criteria	Score (1-10)
Frequency (number of times used per week)	7
Potential Time Savings	2 hours
Steps in the process	12
Steps that require human intervention	2

Next, you can then compare and contrast the different candidates – and see which ones stand out.

Now as for the first few processes to automate, it's usually not the best approach to look at cost savings. Rather, the focus should be on looking at areas that will allow for easier automation, such as only looking at a narrow process within a business segment that has a limited number of steps. It's also helpful if it does not require much approval and help with different departments.

This will accelerate the learning. Essentially, you want to get a quick win – which will motivate the team and allow for more ambitious goals.

Another way to look at what to automate is to consider certain manual tasks, which often apply regardless of the industry. Here's a look:

- Data Migration: Because a company often has a hodgepodge of applications and software, workers will spend much time just transferring information around. It's definitely a big waste of time and susceptible to mistakes. With RPA, it is relatively easy to handle data migration because of the integrations and low-code capabilities.

- Data Updates: What if a customer has a change of address or phone number? How do you update this across different software systems? It can be a pain for workers. But RPA has abilities to create rules to automate this process – and there may not be a need to seek out the assistance from IT.

- Tracking: Employees simply do not have the time to monitor the many changes within a department. Because of this, it's common for problems to slip through the cracks. However, RPA can be configured to detect changes and take action – without human intervention.

- Alerts: You can program a bot to detect when there needs to be a decision from a worker. For the most part, RPA is ideal for this.

Once you have identified the processes, it's a good idea to then have documentation of them. This will help provide rigor and discipline as well as help with getting new team members up to speed. The documentation does not have to be highly detailed either. Rather, just make sure you have:

- Rationales for the process
- Steps or flow charts

- Names of the workers involved

- IT assets involved

- The requirements for security, compliance and governance

OK, what processes are not good for RPA? According to Aaron Bultman, who is the Director of Product for Nintex RPA: "Tasks involving creative thinking, brainstorming, interacting with the physical world – like pulling papers from a filing cabinet – are better handled by people. But just because a process involved something like this, doesn't mean you can't automate the repetitive parts. A workflow automation tool allows you to automate the repetitive stuff and send an assignment to a human worker to complete. Once that task has been completed, the workflow resumes. And in some cases, tasks can be sent off to a RPA bot to be completed on the desktop. You see, when you have both workflow and RPA, you can automate far more in your organization. The idea is to only leave the human work for the human to perform. Nobody likes working on low value, thoughtless or repetitive work."[7]

Something else to consider: You might want to start with back-office functions. Why? These tend to be tedious and rules-based. As for front-office applications, there can be much more variability. What might a customer do on the phone? How might there be interaction with a chatbot?

There is also the risk of having a bad customer experience if the bot is not developed properly.

---

[7]From the author's interview with Aaron Bultman, who is the director of product of Nintex, on October 9, 2019.

ROI for RPA

Getting a sense of the ROI is tough in the early stages of an RPA implementation. But it's important to come up with something. This will help to setup some milestones to achieve and also provide a way to track the overall project. But to get a good measurement of the ROI, you will likely need to wait for at least a year.

What is part of the calculation then? The "return" part of the equation is generally the costs that are saved from the RPA project, such as for the less need of full-time equivalent employees (FTEs).

Example: Suppose you want to automate the invoicing process and it requires four FTEs who are paid $20 per hour. They work on the process 40 hours a week. Thus, the total labor costs will come to $166,400 per year.

As for the RPA system, it has reduced the costs by 60% or to $58,240 per year – which puts the net gain at $108,160 ($166,400 minus $58,240).

Next, the annual costs of the RPA system are $80,000. This means that the ROI is as follows:

Year	Amount Saved	ROI
1	$108,160	35.2%
2	$216,320	170%
3	$324,480	306%

In this example, this project would certainly be a big winner.

Yet when looking at the ROI, you may want to take a broader look. For the return side, you could track such areas like the following:

- Accuracy: What has been the decline in errors compared to handling the process in a manual way? Also, is the data quality better because there are fewer mistakes with data input?

- Customer Satisfaction: Has there been an improvement? One common metric to look at is the Customer Satisfaction Score (CSAT), which is a customer survey that asks "How satisfied were you with your experience?" Or you could use the Net Promoter Score (NPS). This shows a customer's interest in recommending a company's product or service – and the index ranges from –100 to 100.

- Agility: How fast is the process now? After all, a bot can work at high-speed on a 24/7 basis.

- Employee Satisfaction: You can send out a periodic survey to see if there has been an improvement. Furthermore, look to see if absenteeism and turnover rates have dropped.

- Innovation: With the RPA handling mundane tasks, has there been more time for employees to provide higher added value? Are they creating new approaches or innovations?

- Analytics: You can get better tracking of your processes, which can provide even more insights on what areas to improve upon.

You can note some of the qualitative benefits. Has there been more reliability to the process? Is compliance improved overall? With such things, you can include them in a report, which should help with the buy-in for the RPA project.

As for costs, the main line item, of course, is the license or subscription fee for the software. There may also be other IT expenses, such as for purchasing prebuilt bots, servers, hosting services, and other software. Then you will need to account for the labor costs for the implementation, bot development, and monitoring, which may include the retaining of a consulting firm. This is why many companies and consultants have used a more comprehensive measure called total cost of ownership, which looks at indirect costs over the life cycle of the automation as well as direct costs (i.e., software licenses).

The good news is that – over time – there should be economies of scale. In other words, certain line items, such as for training, may decline. But then again, there will also be declines with the return because there will be fewer opportunities for "low hanging fruit" for automation.

RPA Use Cases

As RPA has been around for a while, there are many use cases for different departments and industries. This can certainly provide ideas on what areas to start an implementation. Might as well focus on those places that have worked for others, right?

Definitely.

Let's first look at the use cases based on departments:

Customer Service	Validation of checks
	Customer reminders and notifications
	Processing of customer feedback
	Out-of-hours responses
Finance	Credit approval
	Collections process
	Statement reconciliation
	Invoice creation
	Budgeting
	Accounts payable
	Accounts receivable
	ERP data entry
	Profit and Loss preparation

HR	Talent recruitment
	Employee onboarding
	Termination
	Payroll
	Benefits administration
	Employee training
	Compliance reporting
	Employment history verification
	Expense management and reimbursement
	Time record validation
	HR spend analytics
	Requisition management
IT	Configuration and installation
	Server monitoring
	Batch processing
	Email processing and management
	File management
	Backups
	System queries
	Data migration
	Password reset and unlock
	Data validation
	Security monitoring
	Event management
	Application testing
	Help and service desk management
Marketing	List/contact management
	Social media tracking
	Email campaigns
	CRM updates

(continued)

Procurement	Management of work orders
	Processing of returns
	Vendor administration
	Freight management
	Contract management
Tax	Item preparation
	Posting of accounting items
	Entity management
	Modelling

Next, here's a look at some of the industry-specific areas for RPA:

Banking	Loan processing
	Account opening
	Account closure
	Transaction processing
	Risk management
	Approvals
	Confirmations
	Fraud detection
Insurance	Claim verification
	Appeals
	Commission calculations
	New customer packages
	Policy changes
Telecom	Customer dispute resolution
	Phone number porting
	Credit checks
	Processing customer requests
	Collections

Retail	Call center
	Demand forecast
	Returns
	Promotions/discounts
	Supply chain management
Healthcare	Patient management and administration
	Appointments
	Insurance coding
	Patient data migration

The Plan

So far, we have looked at the ways to understand and prepare for an RPA implementation. As you hone in on the key areas for the project, it will be time to put together a written plan. It does not necessarily have to be long or detailed. Instead, the focus should be on developing a clear and effective road map.

You can start out with the high-level objectives and goals. Suggestions include:

- Reduce the FTE time worked on processes

- Improve cycle and turnaround times

- Lower the number of errors

Again, estimating these will, in most cases, be far from accurate. But it is important to set forth clear-cut milestones to hit. And if they prove too onerous or not particularly challenging, you can then make midcourse adjustments.

The RPA plan should also be consistent with the digital transformation goals of the company. This should also be backed up with a champion or sponsor of the project, hopefully from the executive ranks. This will be an indication of the importance and commitment to the RPA project.

Next, the plan needs to set forth the members of the core team. Often this is a diverse group, including people from different departments. There will also need to be a strong leader since the team members will likely not be working on the RPA project on a full-time basis.

No doubt, the plan must be clear-cut on the roles and tasks for each team member (in chapter 6, we'll look at how to form a CoE, which can serve as the team). And then it's a good idea to involve the IT department in the process so as to make sure that there are no issues with integration, security, and governance.

Then the plan can look at the factors for the RPA vendor evaluation. In Chapter 5, we'll look at this in much more detail.

Finally, the plan should have a project road map that shows timelines and deliverables. Usually, this will begin with the pilot or proof of concept. As noted earlier in the chapter, the first project will be more limited, which should allow for more learning and getting to faster results. But the plan can also cover the longer range goals for the RPA efforts, in terms of areas to automate and departments to cover.

Conclusion

Even before you create your first bot, there is much to do! But these steps will be essential for success. If an RPA project fails in the early stages, it could ultimately doom any chances for the use of the technology. Management may simply just cancel the program.

Then when can you start creating bots? Well, not yet. In the next chapter, we'll take a look at how to evaluate an RPA vendor.

Key Takeaways

- While RPA is an easy setup, this does not mean you should quickly start creating a bot. Instead, a better approach is to do some planning. In the first step, you can have brainstorming sessions or even put together a workshop. It's also important to provide some initial training about the fundamentals of RPA.

- To help with the planning and implementation of an RPA system, many companies do seek the help of a consulting firm. Even though there are upfront costs, the benefits can be significant. A quality consulting firm will have full-time specialists who have seen many RPA systems. There could also be help with other areas like assessing company processes, implementing AI, and even hosting the software. Some firms have created their own technology frameworks and bots, which can help streamline the process. Finally, a consulting firm will have the experience with multi-RPA implementations. This will help with better planning and the avoidance of certain issues.

- Spend time in considering what processes to automate. The fact is that RPA is fairly limited in terms of the use cases. For example, it is really for those processes that are tedious, repetitive, rules-based, and prone to error. It's also good if they are recurring in nature (maybe involving an employee several times a week). On the other hand, if the process requires much judgement, ingenuity, and creativity, then RPA is not a good option.

- When looking at the first processes to automate, it's usually better to look at those that are relatively easy and have fewer steps. The reason is that you want to learn from the experience. Over time, as you get better at creating automations, you can take on more complex processes.

- To evaluate which processes to automate, you can use a diagnostic. You may want to look at only a handful of factors, such as those that involve tedious work, are rules-based, have high volume, and are repetitive. You can then score each and see which ones stand out.

- Some of the tasks that RPA can help automate regardless of the industry include data migration, data updates, and tracking.

- Even early in the project, it is advisable to estimate the ROI. This helps with providing clarity of the objectives and should help with buy-in.

- The "return" in ROI for an RPA project is generally the cost savings, such as in reduced number of FTE hours. But there are other benefits like fewer errors, improved compliance, and higher customer satisfaction.

- The "investment" part of the ROI involves the costs of the software, the employee time, consulting fees, and other software/hardware costs.

- The written plan for the RPA implementation does not have to be long. Instead, think of it as a road map. In the plan, you should provide details about the objectives, roles, and deliverables (with timelines). It is also important to have a champion, from the executive ranks, who will make sure the project remains a priority.

CHAPTER 5

RPA Vendor Evaluation

The Right Factors to Consider

Depending on how you define RPA – which can be somewhat fuzzy – there are over 70 software vendors. One reason for this is that the market is still in the early phases and is evolving, such as with technologies like AI. Of course, with the recent surge of funding in the category, this has incentivized plenty of entrepreneurs to create their own start-ups. Oh, and building RPA software is not necessarily difficult (at least when it involves functionality like screen scraping).

Regardless, the profusion of RPA vendors means that it is unpractical to have a comprehensive evaluation. Rather, you will need to come up with a short list that generally fits your requirements.

This is what we'll cover in this chapter. We'll look at the key factors on making the right selection of an RPA vendor.

Be Realistic

Evaluating software can be tedious and time-consuming. It is also tough to get through the inevitable hype. What is real vs. the fluff? It can be tough to tell.

But the decision requires much care and diligence. And it is not just about the expenses.

© Tom Taulli 2020
T. Taulli, *The Robotic Process Automation Handbook*,
https://doi.org/10.1007/978-1-4842-5729-6_5

"A relationship with an RPA vendor could be akin to a marriage that doesn't end well," said Thomas Phelps, who is the CIO at Laserfiche. "Unfortunately, it's relatively exciting to get into, but very difficult and painful to extricate yourself. Vendor lock-in is inevitable. You can't click a button to download all the process automations you built, and repurpose it on another vendor's platform. Switching costs will be high if you change to another vendor."[1]

What's more, the capabilities of an RPA system can vary widely and many of them will likely not have much relevance to your business. The fact is that there is no perfect solution. In other words, there will be certain areas where you'll need to make compromises.

Check Out Third Parties

A good way to start with creating a short list is to refer to the surveys and research from analyst organizations like Forrester and Gartner. They have the resources and expertise to do deep dives on RPA software.

"Look at Gartner Peer Insights, which are validated user-sourced reviews, to get perspectives from real users—from CIO-level to the enterprise architect," said Phelps. "Peer Insights provides a wealth of high-level information on the overall customer satisfaction with a vendor, vendor evaluation and negotiation process, implementation, technical support and other areas."

Another good source is to look at software review sites. They will have their own analysis and user-generated ratings and comments. True, some of the content may be skewed but you should still get a good sense of the main themes.

---

[1]From the author's interview with Thomas Phelps, who is the CIO at Laserfiche, on October 15, 2019.

Here are sites to consider:

- G2.com (`www.g2.com/categories/robotic-process-automation-rpa`): Since 2012, the site has accumulated 843,500+ reviews. As for its RPA section, the top-rated vendors include UiPath, Automation Anywhere, WinAutomation by Softomotive, and Blue Prism.

- Gartner (`www.gartner.com/reviews/market/robotic-process-automation-software`): The site has the following as the top-rated vendors: UiPath, Automation Anywhere, Blue Prism, and TruBot.

- TrustRadius (`www.trustradius.com/robotic-process-automation-rpa`): Launched in 2014, the site has over 211,000 reviews. The top-ranked RPA software vendors are UiPath, Automation Anywhere, Blue Prism, and WorkFusion

- HFS Research (`www.horsesforsources.com`): The firm was the first to recognize RPA as an emerging technology category in 2012. The research is high quality and much of it is free.

So for the rest of the chapter, we'll take a look at the criteria for making an RPA selection.

Minimum Capabilities

Because of the fervent interest in RPA, some software companies have used the term to describe themselves – even if there is little connection! There is nothing new about this. But of course, it could mean terrible results for the customer.

This is why you want to make sure the software has core RPA functions. At a minimum, this means that there must be the creation and deployment of bots, which carry out the automation. While a key part of this is screen scraping, it is also important to have built-in integrations for common applications. And there should be a robust management console for the bots (such as allowing for tracking, orchestration, and analytics) as well as ways to handle business exceptions (this is when a person needs to intervene because the bot does not work).

The first stage of the evaluation should also include a Proof of Concept (POC). After all, RPA vendors often have trial periods for their software. So why not start a small project to get a sense of the features and capabilities?

This often gets confused with a pilot program. But there are clear differences. A POC usually happens in the early stages and is for a development environment. The pilot, on the other hand, is for a real-world application that will go into production.

Who Is the User?

Yes, this question seems kind of simplistic. But the reality is that – when making the decision to purchase RPA software – there could easily not be enough attention paid to the needs of the end user. All in all, this means there will be a risk of a lack of adoption.

"In evaluating RPA software, make sure you select the right tool for the right audience," said Niraj Patel, who is the managing director of artificial intelligence at DMI. "For example, for front-end RPA, which connects consumers to bots, you will need to select conversational technologies, as opposed to backend RPA which is core system software-oriented. Also, for front-end RPA, you may want to make your call center agents bot owners. Platforms can allow them to measure and monitor bot performance. With backend management, technical personnel should manage and evaluate the quality of the work that is performed."[2]

For the most part, the end user will not be a technical person – instead, he or she will be an administrative or a business person. "There is a trick to evaluating RPA vendors – and it's a major one, because this deviates so much from, say, ERP vendor selection," said David Lee, who is a senior research analyst of applications – Data and Analytics at Info-Tech Research Group. "Active business participation is so critical, and in fact, IT won't be the one testing out the bot creation – the business must be the ones building the bots. This gets missed a lot, because too often businesses view this as another technology solution, and expect IT to take it away and come back with a recommendation. This does not work."[3]

Funding

For the first half of 2019, venture funding remained on a torrid pace, with $62 billion in investments in US start-ups.[4] Assuming the pace continues, it will be an all-time record for the year. Foreign markets have also been strong at about $104 billion.

Yet the funding markets tend to focus on a relatively small number of deals because the investment amounts have increased significantly. In other words, even though a category like RPA may be red hot, it could still be tough for many of the players to get sufficient funding.

---

[2]From the author's interview with Niraj Patel, who is the managing director of artificial intelligence at DMI, on October 2, 2019.

[3]From the author's interview with David Lee, who is a senior research analyst in the area of applications – Data and Analytics at Info-Tech Research Group, on October 10, 2019.

[4]https://techcrunch.com/2019/07/06/startups-weekly-2019-vc-spending-may-eclipse-2018-record/

This is certainly something to note when it comes to evaluating a vendor's software. Will the company have the resources to invest in R&D, infrastructure, and support? Also, can a small start-up compete when it comes to hiring talent? And will there be enough resources to pull off acquisitions?

Ecosystem

Even the large RPA players do not have the resources to build a true end-to-end solution. In most cases, a customer will likely need other software to fill some of the gaps (and some of them can be quite glaring).

Because of this, you want an RPA platform that has a thriving ecosystem of partners and integrations. A clear indication of this is when there is a bot store (i.e., an app store to download apps). This will greatly cut down on the time for development and also should improve overall performance of the bot. According to Automation Anywhere, its Bot Store has realized cost savings of 50% and 70% from faster automation of processes.[5]

Costs

Of course, this is usually the most top-of-mind consideration when it comes to selecting an RPA system. But if there is too much focus on this, you may ultimately buy the wrong platform, which could severely hamper your project. Keep in mind that the pricing among RPA vendors is quite competitive. So ultimately, the main part of the analysis is whether the software can solve your problems. It's really that simple.

---

[5]www.automationanywhere.com/products/botstore

But when it comes to the costs, there are notable minefields. Some platforms require a large amount of consulting for the setup. These costs can be two to three times the amount for the software. For large enterprises, you can easily spend multimillions on consulting.

This is not to imply that the money is wasted. In certain circumstances, there may need to be much reengineering. And this is fine. But the key is that the service providers must be clear about what to expect.

In terms of the costs of the software, here are some things to think about:

- Scale: Many RPA projects are still not widespread across organizations. But what will the costs be if you scale the technology? Will the ROI be enough? To answer these questions, you should model the scenario of a large deployment (and make sure the calculation includes consulting costs and hardware and server requirements, which can escalate as the project scales). Thus, when evaluating all this, you will notice significant discrepancies between the vendors. However, this does not mean you should avoid the vendor if the costs are high. Instead, you could use this as an opportunity to negotiate for discounts based on volume tiers and multi-year agreements.

- Nonhuman RPA Access: This can be a gotcha, which leads to higher future costs. "Some RPA vendors are looking to create new pricing metrics – such as licensing nonhuman access or charging for transactions performed using their system – to recoup potential losses from expected decreases in user-based software and subscription revenue," said Thomas Phelps.

OK then, so is there any way to get a general sense of how pricing works? Well, things can vary, but let's take an example that should give enough context on how things work. With a standard RPA vendor, there could be annual payments for the studio/bot designer and the orchestration system, which could easily be over $20,000. But the main cost will be for the bots. Note that most RPA vendors charge on a per-bot basis and the unattended ones usually are much higher. For example, an attended bot could be $1,000 to $2,000 each, whereas the unattended bot could range from $5,000 to $10,000 each (all these costs are payable each year).

The per-bot pricing can also be far from straightforward. The fact is that bots among the different software platforms are different, which makes it difficult to make true comparisons. As a result, you should ask questions about the pricing – and see what you are really paying for. If anything, this could lead to a discussion for lower pricing, especially if you mention features of bots from rival firms!

But some RPA vendors – especially those that focus more on the business process management (PBM) market – may have other pricing models. One would be for inputs like the number of full-time equivalent (FTE) employees or time and materials. And yes, there is the output model, which is based on the volume of transactions. Although, in the end, the per-bot pricing model is the main approach in the RPA world.

Training and Education

A good way to evaluate an RPA system is to take some of the courses. Is the software easy to use? Does it have the features you need?

Some RPA vendors also have free downloads of the software or community editions. Such options are definitely worth looking at.

Then check to see if there is an online forum. What are the comments? Is there a thriving community? Or has it been a while since anyone participated? Or are there many complaints about the software?

Support

You want an RPA vendor that provides full-on support. This would include self-service options like videos, forums, courses, white papers, documentation, and blogs. Also what is the level of technical support? Can you call 24/7?

Then there is the SLA or service-level agreement, which is a contract that sets forth the level of services the vendor will deliver.

Best-of-Breed vs. End-to-End

The seemingly never-ending debate with business and enterprise software is the difference between best-of-breed and end-to-end solutions. True, the nice thing about an end-to-end solution is that there is usually less training and the costs are lower. But of course, the big drawback is that some of the features will be subpar. Even the world's top technology companies have major deficiencies in their platforms. It's inevitable.

With RPA, it's not realistic to have a true end-to-end solution. This is definitely the case for global enterprises, which have extremely complex needs. It's actually not uncommon for such companies to have several RPA solutions. Yet as noted earlier in the chapter, it's a good idea to look at those platforms that have a rich ecosystem of partners that offer complementary technologies say for OCR, ML, and NLP.

Thought Leadership and Vision

Yes, such things are nebulous. But when it comes to business software, you want to see if the vision is in line with yours. When it is, the results can be very powerful.

To get a sense of the vision, you can check out a company's Twitter page, blog, and YouTube channel (viewing the company conference videos can be particularly helpful).

Interestingly enough, you may realize that – with this process – a company's vision is muddled!

And in some cases, a company's vision may verge on hype. It's as if the technology is too good to be true. If an RPA vendor claims that its software can do just about everything, for a low price, and there is little effort – then be wary! This may just be a way to gin up more PR and sales.

"RPA isn't a magic bullet to do every menial task an employee takes on," said Arvind Jagannath, who is the director of product management at AI Foundry. "It'll take some tweaking and adjusting in the early stages before errors start to disappear. It'll also require an upfront time investment to get acquainted with the tool."

Industry Expertise

Even though many RPA vendors claim that their platforms are a fit for many industries, the reality is usually something much different. There is instead probably concentrations in certain verticals.

So you need to do some research on this.

Security, Monitoring, and Deployment

As we'll go into more detail in Chapter 8, there are several potential security risks with RPA software. This is another reason why IT should be brought into the RPA process early on. By doing so, you can greatly minimize the problems.

A solid RPA system should also have strong monitoring as bots can be brittle. So you want a way to detect the issues.

What's more, how will the software be deployed? Will it require an on-premise setup? Or can you host the software from the cloud?

What Type of RPA?

Do the processes you want to automate require human intervention? Or do they work behind the scenes? Such questions are critical when evaluating RPA software. For example, your needs may require only one type of automation, such as attended or unattended (we discuss these in Chapter 1).

Granted, you might want both attended and unattended bots (which, by the way, is common). Then again, the key is that you understand this early in the RPA implementation process so you will make a better choice on the software.

The Design

One of the keys to Steve Jobs' huge success was an obsession with cutting-edge design. He once said, "Design is a funny word. Some people think design means how it looks. But of course, if you dig deeper, it's really how it works."[6]

---

[6]https://medium.com/cygnis-media/20-inspiring-quotes-about-ux-design-to-get-your-creative-mojo-back-3174f6d6666a

It's a crucial distinction – and something to think about when selecting an RPA solution. Let's face it, all of them will boast that their design is intuitive and easy. Almost magical! But if you try it out, you'll often have a much different experience. The sad fact is that much of enterprise software falls well short when it comes to design.

Because of this, there needs to be some time spent with evaluating the workflow, UI (user interface) and UX (user experience). Is it something that workers can learn quickly? And does the software require technical skills – say, some coding?

Next-Generation Technologies

An RPA company will not provide a detailed look at the product road map. Part of this is due to competitive reasons. But a company also does not want to commit itself to tough deadlines. If there are continual delays, customers will eventually stop listening and could go elsewhere.

Despite all this, there are still certain areas that an RPA company should be focused on. Perhaps the most important is AI or ML. These technologies hold tremendous promise in terms of boosting the impact of automation, such as with handling unstructured data.

"Bots as we know them today are best suited for highly repeatable, structured data," said Bill Galusha, who leads Product Strategy for RPA and Content Intelligence at ABBYY. "But 80%+ of enterprise content is unstructured. What enterprises are requiring for their digital transformation projects is AI technology that complements the bots, making them smarter and able to transform unstructured documents into structured content. This removes process friction and accesses more value from business content. Any type of use case involving content is a prime candidate and many of these processes have customer touchpoints, whether its onboarding a customer, filing an insurance claim, or requesting a loan or line of credit. The concept of a 'skill' is something that resonates with business users. This is where an enterprise using an RPA tool or other automation tool could easily add a skill to a digital worker to give that bot more advanced training and understanding for processing content. We see this growing in popularity as marketplaces become more established and offer hundreds of these potential skills that increasingly make the digital bot smarter."

Conclusion

Making the decision for the purchase of RPA software is definitely a big one. It can be scary. As a reminder, it is a good idea to seek help, say, from a service provider, consultant, or implementation partner.

After all, as seen earlier in the chapter, there are lock-in issues. So when making your evaluation, it's essential to take a long-term approach. Will the vendor have the resources and vision to go where you want?

There's no perfect answer. But at least you will have done some due diligence to lower the risks.

In the next chapter, we'll take a look at another critical part of the RPA process: assembling the CoE (Center of Excellence).

Key Takeaways

- It's daunting to select RPA software. There are 70+ vendors on the market. This is why you need to come up with a short list.

- There will not be a silver bullet. Expect to make compromises with RPA software. You may also need to use other software, such as for OCR or NLP.

- A helpful approach to narrow down the list of vendors is to look at research from Forrester and Gartner. There are also review sites, such as G2.com and TrustRadius.

- An RPA system should have minimum functions, such as a bot builder, a management console, and functions to handle exceptions.

- Venture funding for an RPA vendor is another important factor. This will provide some indication that the company has the wherewithal to be around for the long haul.

- While the cost of RPA software is important, it should not be the deciding factor. Ultimately, you want technology that meets your needs.

- The typical cost for RPA software is to pay a fee on a per-bot basis. There may also be a license for some modules, like the designer. Furthermore, it's always a good idea to ask for discounts.

- With your short list, you can evaluate the software by taking company-created courses. How is the UI and UX?

- Look for a company that provides strong support – such as with blogs, videos, and 24/7 phone assistance.

- Other essential considerations for selecting RPA software include security, monitoring, and deployment options (cloud, on-premise, and hybrid environments).

- And finally, does the RPA vendor have next-generation capabilities, such as AI?

CHAPTER 6

Center of Excellence (CoE)

Assembling an RPA Team

In 2011, the prospects for Adobe looked dim as the growth had stalled. The company's CEO, Shantanu Narayen, knew he had to take bold actions to get things back on track. One of his moves was to transition the business model to subscriptions and the cloud. There were also efforts to unleash more innovation across the organization, such as with its Sneaks program (where employees submit ideas and Adobe then provides resources to commercialize some of them).

But there was something else that was critical for the transformation – that is, a focus on streamlining operations with more automation. To this end, Adobe launched its RPA implementation in the summer of 2018. It was a challenge, though. Keep in mind that the company has more than 21,000 employees across the world.

To help with this, Adobe set up a CoE. In a blog post from The Enterprisers Project, the senior vice president and chief information officer of the company, Cynthia Stoddard, writes: "Adobe is still very much in the growth phase of RPA implementation. We continue to find applications for it. Nearly two years in, I would attest that any CIO or IT leadership must

© Tom Taulli 2020
T. Taulli, *The Robotic Process Automation Handbook*,
https://doi.org/10.1007/978-1-4842-5729-6_6

insist upon a CoE as part of any kind of RPA program. I don't think we would have achieved nearly so much success so quickly without it."[1]

At first, she focused on the finance department, which saw strong results. Consider the following:

- There was a 79% reduction in hours spent on the creation of purchase orders for hardware.

- The RPA system now handles 82% of the work for wire transfer requests.

- There was an 82% reduction in manual hours spent for contract creation.

All in all, the results were standout. And as should be no surprise, RPA is a strategic priority at Adobe.

So in this chapter, we'll look at the CoE and some of the best practices.

What Is the CoE?

According to UiPath: "A Center of Excellence (CoE) is essentially the way to embed RPA deeply and effectively into the organization, and to redistribute accumulated knowledge and resources across future deployments."[2]

This is a good high-level definition. But there are different variations and approaches to the CoE. It could be a small team, say, two or three people – or much larger. Or there could even be a CoE for different divisions in a company or departments.

Regardless of the structure, the main focus should be on the "E" of CoE. In a blog from Automation Anywhere: "All businesses require focus

---

[1]https://enterprisersproject.com/article/2019/10/rpa-robotic-process-automation-how-build-center-excellence

[2]www.uipath.com/rpa/center-of-excellence

and alignment in order for RPA to be implemented successfully. Setting up a CoE will provide leadership, best practices, research, and focus on areas for higher business efficiency. Of course, there are many reasons for implementing RPA, such as compliance, cost reduction, improved service levels, revenue, and error reduction. You may have 200 ideas in the pipeline, but you need a dedicated focus to implement a few of them with automation."[3]

In fact, as an indication of the importance of the CoE, Automation Anywhere has created its own dashboard for this (it is accessible from a mobile app). With it, you can easily see the ROI, time savings, and other key metrics of the RPA system.

Finally, a CoE can be run completely by the company that is implementing the RPA system. Or, in some cases, it could be outsourced to a consulting firm (however, there will likely be members from the client company on the CoE).

Why Have a CoE?

There are really few drawbacks when creating a CoE. Perhaps the most reasonable one is that – if you are mostly testing out RPA technology – there is probably not a need to get formal.

However, if you are committed to making the technology work for the long haul, then a CoE should be a part of the process. It is a clear sign that your organization is serious about RPA. The CoE shows that you are willing to make the investments of time and money. And yes, employees will take notice, which will help with the training and adoption. This is highlighted even more if some or all the members of the CoE are full-time.

---

[3]www.automationanywhere.com/blog/product-insights/getting-started-with-automation-7-steps-to-building-and-managing-bots

But there are definitely other advantages to think about, including the following:

- A CoE provides a way to get a 360-degree view of the RPA project. This makes it easier to get the full value from the technology, as there will not be the problems with silos and multiple RPA systems. There should also be a faster implementation, better creation of bots and more diligent monitoring.

- If the CoE is self-run, then there could be lower costs. But of course, there will be much more control over the RPA project, as your organization will be in a better position to experiment and innovate.

- The early phases of RPA tend to be successful. However, the real issue is when the technology is scaled. But with a CoE, you have a better chance with this because there will be concentrated effort on the RPA system.

- The CoE allows for buy-in. This means getting involvement from the C-suite (e.g., in the example of Adobe, there was collaboration between the CIO and CFO). But there should also be involvement with those on the frontlines who use the technology.

"Implementing a Center of Excellence (CoE) for automation is crucial to sustain RPA success," said Agustin Huerta, who is the VP of Technology, AI, and Process Automation Studios at Globant. "A proper CoE is necessary to serve as the guardian of technical standards and impose a single methodology for approaching automation, so as to ensure that the technology can work most effectively across the organization once implemented. The CoE must also act as the gatekeeper of the processes that will be automated, so as to ensure they have a proper ROI, along with governance of the new digital workforce."[4]

Forming the Team

When it comes to forming the CoE, it is usually in four to six months after the start of the project. This makes sense because the time allows for learnings, such as experimenting with proof of concepts. But during this period, it's a good idea to think about the CoE, such as by identifying the potential members and considering the goals. This will certainly allow for a much smoother transition.

OK then, so what are the different roles for a CoE? Well, in general, you want to get a broad representation of the organization impacted from the RPA implementation.

"Often, the firm will partner with experts to support part of the team who understands how RPA works," said Pat Geary, who is the chief evangelist at Blue Prism. "The subject matter experts (SMEs) work alongside those who understand the automation and process management roles to help them come together and prioritize the most effective automation strategies. The technology team must be a part of the CoE from the beginning as well. While process-based work may not be second nature to the IT team, the technology to support the automation certainly is. By engaging IT teams early, they can gain a better understanding of the platform and can provide additional guidance when configuring digital workers, following unique requirements for security, compliance, and auditing. As IT teams become more familiar with the platform, they often adopt digital workers to support their own processes.

---

[4]From the author's interview with Agustin Huerta, the VP of Technology, AI, and Process Automation Studios at Globant, on November 15, 2019.

This teamwork allows a digital workforce to be more easily integrated into an organization's core fabric."[5]

On a more granular level, you can look to UiPath's own approach to setting up a CoE. Actually, it's quite extensive, involving a long list of roles! However, when starting out on an RPA project, it's recommended to start small and build over time.

Despite this, it is still good to see the potential roles from UiPath (regardless if they are part of the CoE or not). In this model, there is one that we've already mentioned in this chapter: The RPA sponsor. Yes, this is someone from the business side of the organization, such as an executive.

Next, here are some of the others:

- RPA Champion: This is the evangelist. That is, he or she will be the person who gins up excitement for the RPA project, such as by putting together videos, workshops, and blog posts.

- RPA Change Manager: This person is similar to a champion. In other words, he or she will look for ways to get buy-in with the RPA project. A big part of this is through the creation of a communication plan.

- RPA Service Support: This person will be the first line of assistance for the RPA project.

- RPA Infrastructure Engineer: This person helps with the server installation and monitoring. He or she may also assist with the architecture of the RPA implementation. The position requires deep technical skills, such as for network protocols and administration, security, cloud systems, server layouts, database technology, and virtualization (say, with VMWare, Citrix, or Hyper-V).

---

[5]This is from the author's interview with Pat Geary, who is the chief evangelist at Blue Prism, on November 12, 2019.

While all these are important, you will notice – at least when you check out online jobs postings – that the main roles include those for the business analyst, developer, solution architect, and RPA supervisor.

We'll next take a look at each:

Business Analyst

This person manages the duties between the company's SMEs (subject matter experts) and RPA supervisors – who have certain requirements for achieving revenue and cost goals – and the developers. It's a tough job and requires a blend of skills that cross business, process planning, and technology. No doubt, he or she needs a deep understanding of the RPA platform.

Some of the activities for a business analyst include the following:

- Develop documents that set forth clear-cut descriptions of the goals of the project and what needs to be done. One is the process design document, which is often created with the help of the RPA developers.

- Look at ways to improve current company processes. From this, a business analyst will come up with an opportunity assessment document.

- Have a high-level understanding of technologies like programming languages and AI.

- Experience in evaluating different RPA solutions.

- Work with IT for integration, security, and governance.

- Put together test cases to determine if the RPA implementation is hitting the goals.

- Help assemble the CoE.

- Monitor and maintain the bots.

Depending on the size of the RPA team, the business analyst may serve the role as the project manager.

Developer

The developer will be responsible for building the bots. But of course, this will involve many other important responsibilities:

- The identification of opportunities to improve business processes through automation. This may involve understanding concepts like lean Six Sigma.

- The creation of comprehensive process definition documents (PDD) and technical definition documents (TDD). This will require collaboration with the team members.

- The conduct of code reviews to make sure the bots are working properly and are consistent with company goals and guidelines.

- The monitoring and maintenance of bots when they are put into production.

- The focus on finding ways to reuse existing bots and code so as to increase the speed of development.

- The creation of easy-to-understand documentation and training materials.

In terms of background, a developer will usually have a technical education, whether a degree in computer science, engineering, mathematics/statistics, or management information systems. And yes, this person should have a certification with the RPA system. Keep in mind that many of the vendors have their own education programs.

Besides all this, it's helpful for a developer to have some knowledge of the following:

- Scripting languages (like JavaScript and HTML) or computer languages (like C#, Java or .NET).

- Agile development techniques.

- Databases like SQL or NoSQL systems.

- Enterprise applications like Salesforce.com, Oracle, and SAP.

If your plan is to have a sophisticated RPA implementation, then a developer should have an understanding of technologies like AI.

Because of the high demand for technical talent, the compensation for a developer is on the high side. According to ZipRecruiter.com, the average salary is $109,135 per year.[6] Although, in order to reduce the costs, a company may retain a developer on a freelance basis. The rates can range from $70 to $150 an hour.

RPA Solution Architect

There is not a bright line between an RPA developer and RPA solution architect as both require strong technical skills. But there are still some differences. An RPA solution architect is generally for the early stages of the project, such as with designing the core foundation for the technology. A big part of the job is to select the right technologies that are best for the goals of the organization.

---

[6]www.ziprecruiter.com/Jobs/Robotic-Process-Automation-Developer

But here are some other activities:

- Create an infrastructure that will scale over time.

- Look at next-generation technologies like AI.

- Put together models and prototypes.

- Analyze the security risks.

- Look at testing procedures.

For the most part, an RPA solution architect will need to have the ability to collaborate with different departments. He or she will also document the procedures.

RPA Supervisor

This is the point person who manages the RPA system. He or she will be responsible for the overall performance but also for the budget. While the role will not include development, there will still be a need for a good understanding of RPA technology and the use cases. There should also be experience with project management. And this does not have to be with RPA. For example, an RPA supervisor may have a background with other software implementations, such as for CRMs or ERPs.

Here are some other skillsets:

- Understanding of customer requirements.

- Be able to work with compliance and security.

- Have strong communication skills.

- Have a track record on delivering on projects.

- Be able to put together clear-cut plans of action.

- Provide ongoing reports and updates.

- Be able to implement controls and guardrails for monitoring and dealing with exceptions.

What Should a CoE Do?

The phrase "center of excellence" sounds kind of like another name for a committee. This could lead people to believe that it is more of an afterthought, in which much does not get done! As Fred Allen once quipped: "A committee is a group of people who individually can do nothing, but who, as a group, can meet and decide that nothing can be done."[7]

Unfortunately, when it comes to CoEs, the performance can be choppy. In some cases, the group is really a nonfactor.

But as we've seen earlier in this chapter, the CoE has been critical for companies like Adobe. The key is that they understand how to set forth the mission.

Then what should a CoE do? There are various operating models. But there is one from Infosys Consulting, which is divided into two stages.[8] First, there is the automation CoE, which is about the steps for setting up the RPA project. They include:

- Automation Strategy: Here, you will look at the vision and roadmap of the RPA project. What is the business case? What functions will be automated first? What is the target ROI? What are some of the qualitative goals, such as improving innovation and customer service?

---

[7]www.brainyquote.com/quotes/fred_allen_201549?src=t_committee
[8]www.infosysconsultinginsights.com/2018/01/18/automation-centers-of-excellence/

- Technology Strategy: This is more than about selecting an RPA platform. There will be a need to consider bolt-on software, such as for OCR, NLP, and AI. Moreover, a technology strategy should focus on the long term: how to build a solid foundation and scale the technology.

- Governance: This involves setting up a compliance structure that is backed up with roles and responsibilities. However, governance is often neglected – which can lead to major problems with the RPA implementation.

- People Enablement: What are the skillsets required for success? Granted, this often gets too focused on technology. But there should also be consideration for strong project management and change management.

- Operating Model: This is concerned about the organization around RPA. For the most part, this is usually about the business side of things.

- Bot Development and Support: This involves setting up the methods and standards for the automation. There will also be strategies for reusing bots (say, from a bot store) and managing them over time.

Next, there will be the creation of the scalable automation execution engine. In other words, this will involve the recurring activities that the CoE will handle after the RPA system has been implemented:

- Demand Management: This is identification and priority of automation approaches.

- Bot Development: Here you will build, test, and deploy the bots.

- Bot Support: This will include monitoring and handling of exceptions.

- Organizational Change Management: This is a broad topic, which is for getting the organization to fully use the RPA system. We'll cover this in more depth later in the chapter.

And finally, Infosys provides different ways to deploy the CoE within the organization:

- Centralized Model: All the duties are managed by the CoE team. This is particularly useful when an organization is already highly centralized, the RPA implementation is in the early stages, and there are relatively few opportunities for automation. In terms of the advantages of a centralized model, they include easier sharing of knowledge and best practices and lower costs. Then again, it can be difficult to manage, especially for large organizations.

- Decentralized Model: This is where there will be multiple CoEs across the organization. This is helpful if there are unique requirements in the departments. Thus, a centralized model should allow for greater ROI. But of course, it could lead to complex silos, duplication, and higher costs.

- Federated Model: With this, the automation CoE is centralized but the scalable automation execution engine is spread across the organization. This approach is meant to help get the best of both the centralized and decentralized models. Yet it can be difficult to pull off.

Communication

In the comedy show The Office, a major theme in the first season was the threat of a downsizing. There were rumors among the workers – and lots of fear. Some were preparing their resumes.

But the boss, Michael Scott, had no clue what to do. In one scene, he said: "Am I going to tell them? No, I'm not going to tell them. I don't see the point of that. As a doctor, you would not tell a patient if they had cancer."[9]

True, most bosses do not say something like this! But it is not uncommon for them to avoid hot-button issues either. Yet this only makes things worse.

With RPA, it's inevitable that words like "automation" and even "AI" will be scary for workers. Will their jobs be axed? Will they actually be training the bots to replace themselves?

So when it comes to a CoE, a major imperative is how to handle these sensitive discussions.

Then what to do? Something to note is that transparency is paramount. Employees have a pretty good way of detecting when they are being mislead. Basically, you should provide the goals of the RPA project to everyone impacted. And if this means that one of them is to ultimately lower head count, then this should be disclosed.

On the other hand, make sure to talk about the benefits of RPA. Describe how the technology will mean having to do fewer tedious activities. That is, there will be more time to devote to more crucial activities.

This messaging should be repeated. You can also put it in videos, blogs and other educational materials. Another approach is to bring in a guest speaker from a company that has had a successful RPA implementation. Or, if this is not practical, you may instead have a short video testimonial.

---

[9]www.quotes.net/show-quote/61913

Yet temper the enthusiasm. Talk about what is realistic. If the bar is too high, then there will likely be considerable disappointment. As the old adage goes: "Underpromise and over deliver."

What's more, management must be accessible for any questions. In a LinkedIn post from the former CEO of Hewlett-Packard and eBay, Meg Whiteman: "Stop the emails and start talking to your teams. Just letting people know that you're conscious of the challenges, aware of the issues and actively dealing with them matters. At the end of the day, improving communication is a continuous process that depends on individual action – new corporate initiatives and tools will only get you so far."[10]

Change Management

In June 2011, the beleaguered retailer J. C. Penney announced the appointment of Ron Johnson as the company's new CEO. Investors loved the move, as the stock price soared.

Prior to this, Johnson had served for 11 years as the senior vice president of retail at Apple. He had also spent 15 years at Target, where he was key to the merchandising strategy.

As for his move to J. C. Penny, Johnson said the following (in a press release): "I've always dreamed of leading a major retail company as CEO, and I am thrilled to have the opportunity to help J. C. Penney re-imagine what I believe to be the single greatest opportunity in American retailing today, the Department Store."[11] He even invested $50 million in J. C. Penny stock!

But unfortunately, Johnson's actions at the company would prove to be disastrous.

[10]www.linkedin.com/pulse/20130423170055-71744402-the-importance-of-transparent-communication/

[11]https://ir.jcpenney.com/news-events/press-releases/detail/268/j-c-penney-company-names-ron-johnson-as-its-next-chief

He made radical changes to the store design and pricing. He also refocused the merchandising on younger demographics.

The upshot: The loyal customer base stayed away from the stores, as sales plunged. Employees were demoralized and productivity suffered.

It was a classic case study of how "change management" can go awry. For the most part, people do not want *too* much change.

Change management is a vital part of RPA as well. As seen in this chapter, employees will likely be resistant because of the fear of losing their jobs or having to be relocated. But there are other things to note, such as that many companies have legacy systems – perhaps even mainframes. So to be successful, it's important to be mindful of the cultural impact and issues. If anything, change management is often one of the leading causes of RPA failure.

There are a myriad of frameworks for change management. But there is one approach that is often used, which is from author and management consultant John Kotter. He recommends an eight-step process (this was based on research from over 100 organizations and became part of his best-selling book, *Leading Change*):

- Step 1 – Create a Sense of Urgency: There needs to be a major change agent that motivates the organization into action. This can be done by the use of a story or analogy, without jargon and buzz words. There also should not be too much reliance on data. You want to win the hearts and minds of the organization. An example of this is Mark Zuckerberg, when he energized his company to make the transition from web-based technology to mobile in 2012. He made it clear that this was urgent – even existential. He also personally showed his commitment to the vision. "I basically live on my mobile," said Zuckerberg at the TechCrunch conference. "I wrote the founder's letter on my phone.

I do everything on my phone. I still use our web site. But I think it's really clear that you check in more, share more on mobile. A lot of the energy in development is going into mobile more than the desktop."[12]

- Step 2 – Build a Guiding Coalition: Change is more likely to happen if it is powered by influential people within the organization. Yet this is not a list of executives. The power can be based on expertise, credibility, leadership, and reputation. You essentially want to create a wave of momentum for the project.

- Step 3 – Form a Change Vision: Boiling things down, this means there must be a destination. This will greatly help drive real change in an organization. Some of the factors for a successful vision include the following: it is clear, worth achieving, realistic and attainable, and flexible. Kotter writes: "A useful rule of thumb: if you can't communicate the vision to someone in five minutes or less and get a reaction that signifies both understanding and interest, you are not yet done with this phase of the transformation process."[13]

---

[12]www.forbes.com/sites/roberthof/2012/09/11/mark-zuckerberg-we-burnt-two-years-betting-on-mobile-web-vs-apps/#52a69e04ba6f

[13]https://hbr.org/1995/05/leading-change-why-transformation-efforts-fail-2

- Step 4 – Communicating the Vision: It's common to create a vision and then not do much with it! But great leaders are constantly reinforcing the message – to the point where it seems almost too much (Kotter actually advises to amply the message by a factor of ten). But it takes time for an organization to absorb something new. You should also not assume that the employees are clear on what's happening. Patience is key.

- Step 5 – Empower Broad-Based Action: There needs to be a focus on making sure people are recognized for their efforts. This should be the case even if the results do not meet expectations. After all, there should be a willingness to allow experimentation, which means there will be some failure. The main constraint is that the vision must be the focus. What's more, the leaders in the organization should take actions to reduce the barriers and friction in making the change happen. "In some cases, the elephant is in the person's head, and the challenge is to convince the individual that no external obstacle exists," writes Kotter. "But in most cases, the blockers are very real." But there are other obstacles to note, such as the compensation structure and the reliance on employee titles.

- Step 6 – Generating Short-Term Wins: Who doesn't like a big win? But in the early phases of change management, this is not realistic. It can easily take a couple years for there to be real transformation. This could drag on the organization and lead to more resistance. This is why it's better to look at short-term wins, which should help build momentum.

- Step 7 – Consolidating Gains and Producing More Change: It's good to celebrate a win. But be careful. Do not give the impression that it is the end of the project. Rather, make it clear that this is one step in a journey – and then reinforce the long-term stretch goals.

- Step 8 – Anchoring New Approaches in the Culture: Sustaining the change is perhaps the hardest part of change management. As much as possible, you want to get to the point where people say: "This is the way we do things around here." Basically, the change has become a permanent part of the culture. Kotter says that this is possible when the following are done: there are clear examples of how the change improved the organization (this requires lots of communication) and allowing enough time to make sure new managers can adopt the approaches.

Kotter's methodology is extensive and is far from easy to pull off. But he has demonstrated that it has led to great outcomes.

But with RPA, there are some other factors that can help as well. Here's a look:

- Gauge the Sentiment: It can be tough to understand if the change is taking root. Do the employees really believe in it? To help with this, you can have periodic surveys (say, with SurveyMonkey) to see how things are progressing.

- Gamification: As the name implies, this is about using methods that mimic a game. This could be having a score for a project or even having contests – say, prizes for the most creative bots! By applying gamification, you can make the process more engaging and productive.

- Strategy: The case of J. C. Penny – while notable –
  is not an exception. The fact is that many change
  management efforts fail. But even using Kotter's
  methodology has its issues. Why? The reason is that the
  process really does not matter if the strategic goals are
  not the right ones. This is why there should be lots of
  analysis and thought on whether to pull the trigger on
  an initiative. This certainly goes for RPA. Even though
  many companies are doing it right now, this does not
  mean you should too. Your organization might simply
  not be ready yet. But interestingly enough, even if you
  select the right strategy, there are still land mines!
  For example, the management team will have a solid
  understanding of the plan. But the temptation is to
  implement too quickly. Consider that it will take some
  time to educate the organization on why certain actions
  should be taken.

CoE Case Study: Intuit

Intuit is the developer of some of the world's best-selling software,
including Mint, TurboTax, and QuickBooks. The company generates $6
billion in revenues and has $1.5 billion in operating income.

In 2018, Intuit initiated a pilot program for RPA with the
implementation of UiPath. The focus was on the finance department,
which had 450 full-time employees. Around 150 to 200 would be involved
in some part of the RPA effort.

Here are some of the results (in Table 6-1):

Table 6-1. *Intuit's outcomes for its RPA implementation*

Bot	Process	Hours Saved	Turnaround Time
Charge backs offsets	A bot verifies that the chargebacks are in the right system.	745	1 day to on demand
Manual invoice Mail merge	The bot conducts a mail merge that has individual invoices attached to the corresponding database record.	702	1 day to on demand
Sales tax exempt	The bot will release the orders with the proper tax exemption received from the database records.	715	4 days to on demand
Credit limit and order placement	The bot sets the credit limit for customer accounts, then releases the quote resulting in the order being placed correctly on credit.	580	1 day to on demand
End-user retail returns	The bot turns off the license for the returned product and processes a suspend order in the Intuit system.	450	10 days to on demand

The leaders of this RPA effort included Mark Flournoy, the VP corporate controller and chief accounting officer, and Maaly Mohamed, the finance automation CoE leader. From the start, they wanted to pursue a digital transformation where the employees could have more time for innovation.

"Digital transformation is as much as a mindset and cultural shift as it is a business and technical shift," said Flournoy. "This includes applying this mindset to our operating model, and finding ways to efficiently scale our team and capabilities, whether this is through partnerships or new ways of operating, like RPA."[14]

It definitely helped that there was cooperation with the technology teams from Day 1, which allowed for building a solid infrastructure. There was also strong executive sponsorship.

And yes, there was the CoE. In fact, Flournoy and Mohamed set forth guiding principles for this:

- Driving digital transformation strategy for now, short and long term, in partnership with the technology teams.

- Offering a variety of automation options to meet business requirements, including RPA as one option.

- Educating that process simplification always comes before automation. Don't automate broken or bad processes.

- Minimizing tech debt, which is not reusable technology that becomes obsolete.

- Following the standard software development life cycle methodology.

- Ensuring that information security and controls are maintained to Intuit's standards. There is no greater risk than human error.

In terms of training, two employees were certified on UiPath. But this was just the start. There are actually 25 signed up for this.

So then what does the CoE look like? According to Mohamed: "We have a team of four, which consists of a leader, an analyst and two developers. The CoE leader is responsible for automation program

---

[14]From the author's interview with Mark Flournoy on December 3, 2019.

management, strategy and execution, program change management and team management. The RPA analyst is responsible for roadmap and backlog management, business process assessment, prioritizing opportunities, reporting and tracking ROI. The RPA developers are responsible for automation building, development for process design, testing and deployment to production as well as ongoing support, monitoring and enhancements."[15]

Something that's very interesting about Intuit's approach is how it onboards bots. It's actually similar to the process of bringing on a human employee!

"The CoE will submit a request for a bot," said Mohamed. "Then a Digital Worker account is provisioned in our HR & Identity Management Systems so that the RPA platform can programmatically request credentials. Once the bots are created and provisioned into requested Intuit Financial Systems and ready to execute their assigned tasks, credentials are stored securely and are updated after every login. Once this process is no longer needed, the bot will be de-provisioned and reassigned to a new set of credentials following the same process. This will leverage the existing access management controls that we use today for humans."

Conclusion

According to the senior product manager at IBM Watson IoT, Heena Purohit: "Over time, the CoE can help create a repeatable process for any team that wants to embark on their own RPA journey and can provide expertise and knowledge sharing for all aspects of an RPA project from project initiation and readiness assessment to delivery and ongoing support. The CoE team can act as 'in-house consultants' to help other teams realize value through automation."

---

[15]This is from the author's interview with Maaly Mohamed on December 3, 2019.

CHAPTER 6   CENTER OF EXCELLENCE (COE)

This is a good way of looking at it. The CoE is a combination of management and change agent – which should lead to ongoing strong results.

Now, as for the next chapter, we'll take a look at the bot development process.

Key Takeaways

- A CoE is a group that helps implement, deploy, and manage an RPA system. It can be a small group, say, with a couple people, or large. It may also be run completely internally or outsourced to a consulting firm.

- Reasons for creating a CoE include getting a 360-degree view of the RPA project, improving governance and security, building more effective bots, getting higher ROI, and gaining buy-in from the organization (this is especially the case if there is involvement from executives).

- The RPA sponsor is someone from the business side of the organization, such as an executive. This person is often crucial for the commitment to the project.

- The RPA champion is the evangelist for the project. He or she will help with videos, workshops, blog posts, and so on.

- The RPA change manager will look for ways to get buy-in for the RPA project.

- The RPA service support person helps with the first line of assistance for the RPA project

- The RPA infrastructure engineer's role is to manage the server installation and monitoring.

- A business analyst manages the duties between the company's SMEs, RPA supervisors, and developers.

- The RPA developer will design the bots but also help with deployment and monitoring.

- The RPA solution architect generally helps with the early stages of the RPA project, say, with the design of the core technology foundation.

- The RPA supervisor manages the team players for the project.

- Infosys has come up with a structure of a CoE that has two main stages: the automation CoE (which is for the setup) and the scalable automation execution engine (which is focused on recurring activities).

- In the automation CoE, some of the main functions include the automation strategy (what is the business case?), technology strategy, governance, people enablement (what skillsets are required for success?), the operating model, and bot development and support.

- As for the scalable automation execution engine, some of the duties are to handle the ongoing bot development and deployment, demand management (what should be prioritized?), and organizational change management.

- The Infosys model also shows the different approaches for deploying the CoE: centralized (all duties are managed by the team), decentralized (there are separate CoEs in the organization, such as for each department), and federated (a hybrid of the first two).

- With the CoE, communication is critical. But there needs to be transparency about the goals. Do not try to avoid certain issues, such as potential job loss. It is also important to highlight the benefits of the RPA project. This will help to encourage higher adoption.

- Change management is essential for RPA success. It's about changing the culture by making automation a priority.

- There are different change management frameworks but there is one that is one of the most used: Kotter's 8-step strategy. It recommends to create a sense of urgency, build a guiding coalition, form a change vision, communicate the vision, empower broad-based action, generate short-term wins, consolidate gains and produce more change, and anchor new approaches in the culture.

CHAPTER 7

Bot Development

A Step-by-Step Approach to Success

Developing a basic bot does not require programming experience. But there is still a learning curve. And if you really want to push the limits, you will need to have an understanding of such things as APIs, VB.net, and other technical tools.

In this chapter, we'll look at how to create a bot by using the UiPath platform. There are several reasons why, such as the following:

- UiPath is one of the most popular systems on the market.

- There is a free version of the software.

- UiPath has a modern interface that is relatively easy to use.

- There are helpful features like search, debugging, and monitoring of bots.

- UiPath comes with hundreds of prebuilt actions, which helps to streamline the bot development process.

© Tom Taulli 2020
T. Taulli, *The Robotic Process Automation Handbook*,
https://doi.org/10.1007/978-1-4842-5729-6_7

- The software can record actions across many environments like the desktop, web, terminals, and Citrix applications.

- There are a large number of integrations, such as with Google and Microsoft.

This is not to imply that you should use UiPath. As we saw in chapter 5, there are a myriad of factors to consider when making a selection.

Again, in this chapter, the use of UiPath is really for demonstration purposes.

Let's get started.

Preliminaries

Compared to the rest of the chapters in the book, this one is challenging. This is especially the case for those without a technical background. Thus, when going through this chapter, do not rush things. I would also install the UiPath software and follow along, which will really help to reinforce the concepts.

In this chapter, we will start with the core elements of bot design, such as creating a workflow or flow chart of the functions. We'll then get to more advanced topics – involving programming – like variables, conditionals, and loops. Finally, we'll cover some of the basics of debugging.

What if you do not plan to use UiPath? Well, in this case, you should definitely check out the vendor's educational materials.

But for the most part, they will likely be similar to what is in this chapter. So even scanning some of this chapter can be helpful in getting an understanding of what it takes to create a bot.

Installation of UiPath

There are two main versions of UiPath. There is the Enterprise platform, which requires the payment of a license (based on factors like the number of bots used). However, you can evaluate it for free for up to 60 days.

Then there is the Community edition, which is free. You also get access to support on the forum. So yes, in this chapter, we will highlight this version.

Yet UiPath does have some rules with the Community edition. An individual or a small company (with a limit of five machines) can use the Studio for business purposes. But this is not the case with the Orchestrator. This is only for evaluation and training.

And if the company is considered an enterprise, then both the Studio and Orchestrator will be for evaluation and training purposes.

As for the installation, go to the following: `https://www.uipath.com/ start-trial`. Then select "Try It" for the Community Cloud version and create a login with either your e-mail or Google, Microsoft, or LinkedIn credential. You will then fill out some information, such as your country and company name.

After this, the portal screen will pop up, as seen in Figure 7-1.

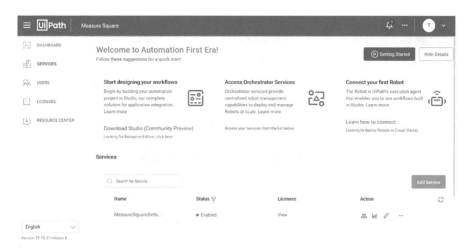

Figure 7-1. *This is the portal screen for UiPath*

A good first step is to click the "Getting Started" button, which will provide a three-minute video on how to use of the core features of UiPath.

Next, I would then install the UiPath Studio (this is where you will create your bots). It is available for a Windows desktop and the download will take a few minutes to complete.

When you launch the application, you will get the following screen (Figure 7-2):

Figure 7-2. *This is the initial screen for the UiPath Studio*

Let's take a look at the various options available:

- Open: You can navigate your computer system to select a UiPath process or select one you've already saved. The "Clone or Check Out" is an advanced feature, which allows you to use version control (say with Github). This means you can track the development of the process with comments and other information. This can be particularly useful when working in collaboration with others.

- New Project: Here, you can create a new process completely from scratch or you can make one for later reuse (it will become part of a library).

- New Template: These are prebuilt processes. If one meets your needs, you can speed up the development process. For example, there is a template that has a flow chart. Then there are more advanced ones, such as for developing enterprise-grade bots.

- Help: This is comprehensive, including product documentation, community forum, help center, academy, and release notes. Also, when you click the "Help" button, you will see details on the right side of the screen. One is the activation ID (this is helpful if you want to upgrade) and the "Update Channel," which means UiPath Studio will either be updated with the stable release or betas (which are not as reliable and may have bugs).

- Settings: You have functions like auto backup and the changing of the UI theme (either light or dark).

- Tools: Here, you can install functions to allow the use of different browsers, like Chrome, Edge, and Firefox. There are also extensions for Java, Citrix, and advanced Windows functions.

Getting Started

In the initial screen of the UiPath Studio, select "Process." You'll see a screen that asks for the name of the process, the directory to put the files in, and the description. As for the name, you cannot use spaces and the convention is to use Pascal case. This means that you will capitalize the first letters of each word. Example: HelloWorld.

Press "Create" and you will get the screen to design the bot (Figure 7-3):

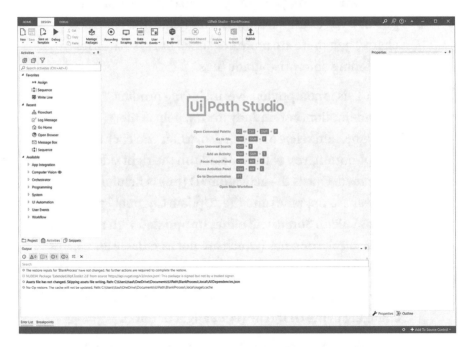

Figure 7-3. *This is the design screen for the UiPath Studio*

At the top of the screen, there are three menu items. Home brings you back to the initial screen. You can then toggle between the Design and Debug screens.

On the left side of the screen, there is a panel that has three tabs: Project, Activities and Snippets (these are small pieces of functions, such as to delay a process or create a loop). The Project will show the main structure of the process and the dependencies, which are the functions that have been included. There will be a small number as a default, such as for Excel, mail, basic automations, and system tasks. If you want more, then you will select "Manage Packages" (at the top of the screen). Here you will have the following options:

- Official: These are the activities that UiPath has created. Some examples include functions for PDFs, forms, databases, and so on.

- Microsoft Visual Studio Offline Packages: With these, you will get added capabilities for coding, such as with the C# language and the .Net platform.

- Nuget.org: Here you will see a diverse set of packages for things like JQuery, Bootstrap, and MongoDB.

When you add a package, it will automatically show up in the dependencies. It will also appear on the Activities tab. This is where you can drag and drop functions onto the center of the screen. But first you must select "Open Main Workflow." By doing this, the middle of the screen will have a canvas that's called Main and it will be here that you will drag and drop the activities.

OK, let's first create a basic bot. Go to the Activities tab and select Sequence. You can do this by entering this in the search box above or click Workflow ➤ Control ➤ Sequence. As the name implies, the Sequence is a placeholder for other activities.

Select this and drag it to the Main screen. Next, you will drag the Message Box activity onto the Sequence (it is located at System ➤ Dialog ➤ Message Box). Where it says "Text must be quoted," you will enter the following: "Hello World!" (make sure this is surrounded in quotes). Here it is in Figure 7-4:

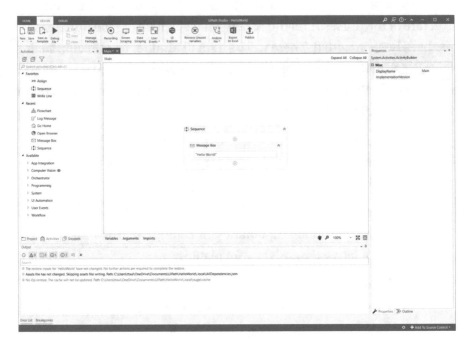

Figure 7-4. *Here's a bot that prints "Hello World!"*

To run this, either click the run button on the menu at the top or press F5.

That's it! You've created your first bot! Granted, it is very simple. But this does provide a general look at the workflow.

With the sequence, you can drag and drop other activities on it. Or, you can insert them by pressing the + button. When you do this, a special menu will appear that will allow you to select the activity, as seen in Figure 7-5. There will be a search box and a list of activities filtered by whether they are your favorites, been used recently, and so on.

Figure 7-5. *Here's the pop-up menu for when you press + in the Sequence*

On the Sequence, you can click the top of the box and you can then rename it. This can be extremely helpful in making your workflows more understandable.

And you can click the icon on the top right to collapse the box. Again, this can be useful when getting a sense of the organization (the activities can get quite busy!). At the top right of the Main screen, you can also select Expand All or Collapse All. Then at the bottom left, there are other helpful buttons to manage your screen (Fit to Screen, Overview, and Zoom).

Another good feature is Properties. By clicking the Sequence or an activity, you will see the box on the right side of the screen that shows the details about the configuration. For example, with the Message Box, you can change the text, the name of the box, the type of button, etc.

Figure 7-6 highlights all this:

Figure 7-6. *If you click within the sequence or activity, you will see the properties in the panel on the right side of the screen*

Activities

When creating bots, you will spend much time in the Design part of the UiPath Studio and work quite a bit with the Activities. Some of these will involve simple drag and drops with a little configuration. But others will be more involved and intricate.

So to get a sense of the types of things you can do with Activities, here's a look:

- Workflow: When creating a bot, you will usually start here, say, with a flowchart or sequence. With the Sequence, there are a myriad of programming-type constructs that help with the logic. Just some include Do While, For Each, If, Break and Continue, just to name a few.

- UI Automation: Here are the main types of automations for OCR (screen scraping and extraction), text, windows, image management, elements (like the mouse and keyboard), and the browser. For example, when you select browser, you can navigate across web pages and even include JavaScript. Or, with a window, you can do things like hide it, maximize it, and restore it.

- User Events: You can monitor for keypresses or other triggers.

- Programming: This involves more sophisticated coding functions. You can engage in database actions, debugging, and setting of variables.

- Orchestrator: This is where you can manage and schedule bots. You can also work with setting up credentials for access.

- App Integration: This is a major area and UiPath Studio comes with a handful of integrations like CSV, Excel, and mail.

- Computer Vision: Here you will find powerful AI features to help with identification of text and other items on a screen.

- System: This shows how extensive UiPath works with a desktop system. You can create/copy/append/move files, detect if a path or file exists, work with the clipboard, open and close applications, manage environment variables and passwords, and invoke PowerShell scripts.

Flowcharts and Sequences

It's recommended to start the creation of a bot with the flowchart or a sequence. The reason? Both are a great way to provide a structure and visualization, which will go from the top of the screen to the bottom.

Since we've already taken a look at Sequences, let's now see how the flowchart activity works. Go to the Activities tab and go to Workflow ➤ Flowchart ➤ Flowchart and then drag and drop it onto the Main screen and click it. You will then see Figure 7-7:

Figure 7-7. *This is the start of a flowchart activity*

Then, as like we've seen with the Sequence, you can drag and drop activities onto the Flowchart.

But they will not run automatically; rather, you will need to create the types of relationships between the activities.

So let's drag-and-drop the following:

- The Open Browser activity (UI Automation ➤ Browser ➤ Browser)

- The Go Home activity (UI Automation ➤ Browser ➤ Go Home)

Now let's connect them. First, right click the Open Browser activity and select "Set As Start Node." By doing this, UiPath Studio puts an arrow from the Start to the Open Browser activity.

Or you can hover over the Start button and you'll see four small boxes on every edge of the button. You can then drag one of these to the Open Browser (or any activity) and the arrow will show up and make the connection. As you will see when in UiPath, there are often several ways of doing the same thing.

Next, drag the bottom box on the Open Browser Activity and connect the arrow to the Go Home activity.

This is what you will have (Figure 7-8):

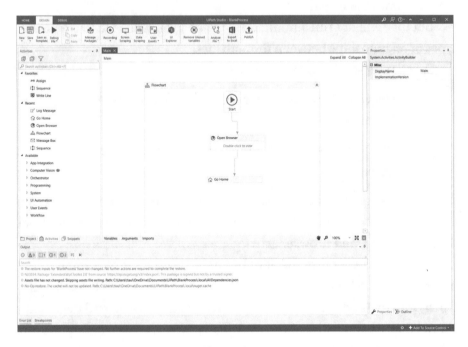

Figure 7-8.   *This is a look at a flowchart*

If you select Start or F5, you will get an error, which says "The workflow has validation errors. Review and resolve them first." In fact, UiPath Studio already provided a warning, with the "!" point on the top-right of Open Browser activity.

The problem? If you click the !, it will say that you need to provide a URL. So put in something like "`www.google.com`" and then start the bot.

From here, you can actually put in sequences or even more flowcharts! In other words, UiPath is highly flexible. But it is important to try to not get too convoluted with the workflows.

Log Message

As a bot gets more advanced, you want to take steps to document the workflow. One approach with UiPath Studio is to use the Log Message activity (it's at Programming ➤ Debut ➤ Log Message). Drag and drop this onto a flowchart or sequence and then enter a comment. You will also indicate the type or log level: Fatal, Error, Warn, Info, and Trace.

When you run the bot, the Log Messages will not show up on your Main screen. Instead, they will appear on the Output screen at the bottom and will be included in the UiPath Orchestrator. There will also be an indication at the top as to the log level (see Figure 7-9):

Figure 7-9. *The output screen shows the Log Message*

Variables

To use RPA, you need to have a good understanding of variables. They are containers – which you give a name – that hold data, such as text and numbers.

To create a variable in UiPath Studio, you will click the Variables tab, which is at the bottom of the Main screen. Select "Create variable" and then enter a name. There should not be any spaces and it's recommended to use Pascal case.

Next you will select the variable type, which include the following options:

- Boolean: This has only two values – that is, true and false. This is usually for loops and if/then/else statements.

- Int32: This variable can only hold integers, which means there are no decimal points.

- String: This is text data, such as contact information, articles, etc.

- Object: This is a custom-built data type, which may include a blend of other variable types. For example, you could create an object that holds contact information.

- System.Data.DataTable: This is to handle large datasets. This is actually common for RPA since it deals with such things as Excel spreadsheets.

- Array of [T]: An array is a string of values. You can use loops to manipulate this type of data.

- Browse for Types: This will open a new menu that provides a large assortment of other unique variable types. This allows you to access the .Net framework, which makes it possible to access deeper into your computer's system. Just some of the types include decimal, date, and password.

Then what does Scope mean? It describes in which part of the workflow you can use the variable. To understand this, let's take an example. Suppose you have a bot that has a Sequence that includes two activities: an Input Dialog and Log Message activity. You also have a string variable called FirstName that you set for the scope of Sequence. This means you can use it anywhere in the Sequence, including the two activities. However, if you set the scope to the Input Dialog, the variable will not be available in the Sequence.

Why do this? It's really about better management of your coding. If you have extensive scope for many of your variables, it can be tough to keep track of things and debug the code. The general rule of thumb is to keep scope as narrow as possible.

What's more, you can set a default value for the variable. And if you do not do so, it will be set to null.

Here's what the Variables tab looks like for our FirstName Variable (in Figure 7-10):

Figure 7-10. *Here is the creation of a new variable called FirstName*

In some cases, you may have a large number of variables. To help navigate this, you can click the top of the column to sort them (by name, variable type, and scope). If you want to delete a variable, click the row and just press Delete. By highlighting a row, you can also create an annotation, which is a description for the variable. This can be helpful when working with a group of people so everyone understands the purpose for each of the variables.

Oh, and there is another way to create a variable. Click a Sequence, Flowchart or activity and press the right mouse button. Then choose Create Variable.

When working with variables in UiPath, there is a useful activity called Assign, which can both create and manipulate variables within the Main screen workflow. To see, let's take an example. Drag and drop a Sequence and place the Assign activity in it (Workflow ➤ Control Flow ➤ Assign). In the "To" box, right-click inside and select "Create Variable." Select this and enter "FirstName" and in the "Enter a VB expression," enter "Tom."

Then click the "Variables" tab at the bottom of Main and you will see the FirstName variable show up. Select "String" for the "Variable type."

Next, drag and drop a Log Message activity on the Sequence (Programming ➤ Debug ➤ Log Message) and chose "Info" for the Log level and type out "The first name is " + FirstName.

Press the Start button or F5 and the message will pop up in the Output box, which will refer to the FirstName variable, as seen in Figure 7-11:

Figure 7-11. *This shows the use of the Assign activity*

Loops and Conditionals

Loops are a core part of computer programming languages. It is a way to repeat a set of instructions – say, to search for something and add numbers – until a condition is met.

The types of loops in UiPath include

- For Each

- Do While

- Switch

Then there are conditionals, which are pathways for the underlying logic of computer code. Even with fairly simple constructs, it is possible to create intricate applications.

In the next few sections, we'll take a look at For Each and Do While/While loops as well as the If/Then/Else and Switch conditionals.

For Each Loop

The For Each loop will cycle through a set of data, such as arrays, lists, database tables, and files. To see how this works, you will first drag and drop a Sequence on the Main screen and then put a For Each activity on it (this will be at Workflow ➤ Control ➤ For Each). This is what you will see (Figure 7-12):

Figure 7-12. *The structure for a For Each loop*

Thus, this will loop through each of the items in the data that is specified in the right box at the top (where it says "Enter a VB expression"). Each time this happens, the instructions in the Body will be executed. Click the For Each box and take a look at the Properties on the right. In the TypeArgument area, you can choose the type of data you want to cycle through.

Let's take an example. In the Properties section, specify String for the TypeArgument and then click Values. Here enter the following:

```
{"John","Jane","Lori", "Ben", "Mary"}
```

Note   If you click the part with the three periods, you will be taken to a larger input box.

Next, drag a Log Message activity to the Body and include the following in the Message area:

```
"Name: " + item
```

Item is a variable that stores each of the names as the loop is repeated.

Press the Start button or F5 and you will see the following in Figure 7-13:

Figure 7-13. *This is the output for the For Each loop*

As you can see in the Output box below, the For Each loop prints out all the names.

Do While Loop and While Loop

A Do While Loop will continue until a condition is triggered. Unlike a For Each loop, you do not know how many times it will cycle through or if it will even need to do so.

Here's an example. Drag and drop a Sequence on the Main screen and put the Do While Loop activity inside it (at Workflow ➤ Control ➤ Do While). In the Body, drag and drop the Assign activity (at Workflow ➤ Control ➤ Assign).

Next, go to the bottom of Main and select Variables and click Create variable. From here, enter the "Counter" for the "Name" and chose int32 for the Variable type. We do this since we want to track the value of Counter as it goes through the loop.

Then go to where it says "To" and enter "Counter" and where it says "Enter a VB expression," you will enter "Counter + 1." In other words, every time there is a loop, the value of one will be added to Counter.

Then drag and drop a Log Message Activity below the Assign box and select Warn for the Log level and put the following for the Message:

"The Count is " + Counter.ToString

You use the .ToString to convert the integer to a string. If not, you will get an error since UiPath Studio does not allow you to mix different data types.

Finally, you will add the following for Condition:

```
Counter < 10
```

Figure 7-14 shows what this all looks like after you press Start or F5:

Figure 7-14. *This shows the results of the Do While loop*

As you can see, the Do While loop counts the numbers until it reaches 10. After this, the loop is terminated.

Then how is the While Loop different? It's essentially the same thing but the condition is at the top of the sequence. This means that the loop may not necessarily be executed. Thus, in our example, this would be the case if the value of Counter is 11 when the While loop is executed.

IF/THEN/ELSE Conditionals

Here's what the IF/THEN/ELSE conditional looks like in the BASIC computer language:

```
IF index > 0 THEN
PRINT "Positive"
ELSE
PRINT "NEGATIVE"
```

Yes, this is self-explanatory. You can also use the IF/THEN/ELSE approach with bots.

Here's how. First you will drag and drop a Sequence and then create a variable called "InputNumber," which will be an int32. Then you will put an Input Dialog activity inside the Sequence (at System ➤ Dialog ➤ Input Dialog). In the Title, enter "Number Box" and in Label put "Enter a number."

You will then drag and drop an If activity (Workflow ➤ Control ➤ If) and then enter the following:

- Condition: InputNumber <= 10

- Then: Drag and drop a Log Message (Programming ➤ Debut ➤ Log Message) and select Info for the Log Level and enter this for the Message: "The number entered is 10 or less."

- Else: Like what we did previously, place a Log Message, select Info for Log Level, and enter this message: "The number entered is more than 10."

Press the Start button or F5 and this is what you will get (Figure 7-15) if you enter 5. That is, the Output screen will show "The number entered is 10 or less."

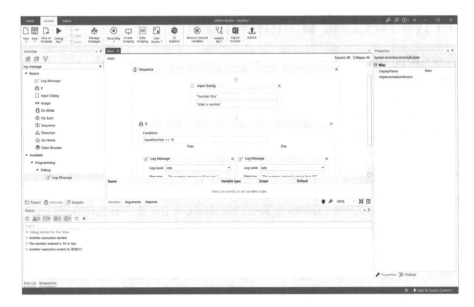

Figure 7-15. *This is an example of using the If activity*

Switch

While the If activity is quite useful, there are certainly drawbacks. Perhaps the biggest is when you have multiple possible pathways – say, more than three. Using the If activity is usually too cumbersome.

But there is an alternative: the Switch statement.

To see how this works, let's go through an example. Create a variable called ChatInput and you will classify it as a String.

Then drag and drop a Sequence on the Main screen and then put the Input Dialog activity inside it (System ➤ Dialog ➤ Input Dialog). In this box, you will do the following:

- Title: Enter "Chatbot"

- Label: Enter "What do you want to do?"

- Result (this is in the Properties Box): Enter ChatInput

Next, you will drag and drop the Switch activity on the Sequence (Workflow ➤ Control ➤ Switch).

Click the Switch box and then change the TypeArgument on the Properties box to String. In Expression, enter "ChatInput." This means that after the user enters text in the Input Dialog, it will be evaluated.

OK, in this chatbot app we are creating, it can only understand three words: Help, Hello, and Purchase! These will be put in separate case statements.

For example, you will click "Add a new case" and then enter Help (note that you do not use quotes). You do this also for Hello and Purchase.

For each of these, you will click "Add an activity" (including one for Default) and then drag and drop a Log Message. For all you will select Info for the Log Level and then the following:

- Help: "How can I help you?"

- Hello: "Hi, how are you doing?"

- Purchase: "What do you want to buy?"

- Default: "I cannot understand the question."

Press Start or F5 and if you enter Hello, you will see "Hi, how are you doing?" in the Output box (see Figure 7-16).

Figure 7-16. *This is an example of using the Switch activity*

Note UiPath also has the Flow Switch activity. This is essentially the same as Switch except it is used for a Flowchart.

Debug

Coding a bot can be fairly quick. What really takes time? It's usually the debugging. In some cases, it can take quite a while. According to the late computer scientist Alan Jay Perlis: "There are two ways to write error-free programs; only the third one works."[1]

---

[1]www.journaldev.com/240/my-25-favorite-programming-quotes-that-are-funny-too

The good news is that RPA systems like UiPath have extensive documentation and helpful forums. A quick search on Google can also provide some links to solutions.

As noted earlier in the chapter, the UiPath platform has a Debug section, which you click on the top of the screen. This provides an array of tools that should help fix problems with the bot.

One is to create more detailed log files, which show the instructions executed, values impacted, and time stamps. This can help pinpoint the nagging issues. To allow for this, you will click the Log Activities. Then you will run the bot by either clicking on the Debug File button or pressing F9. After this, you can find the log file by clicking on Open Logs.

You'll notice that this process can be quick. So by clicking the Pick Slow Step icon and selecting Highlight Activities, the speed will be much more reasonable and you will also see each element highlighted (by the color yellow).

The UiPath Debug system has the ability to set Breakpoints within the workflow. You can do this where you think there may be a problem or before where the bot triggers an error.

Once you reach the Breakpoint, you have a couple options. One is to click the Locals tab, which will show you the contents of the variables. This will indicate if the bot is taking the right actions with the data.

Also, at a Breakpoint, you can then press the Step Into button. This will execute the next step in the workflow – and you can then see how things change. By doing this, you should get a sense of what is happening and where the problems are. Or you can click the Step Over button, which will take you to the next activity.

Another way of debugging a bot is to use exception handling – that is, with the Try/Catch construct. To see how it works, let's first try by dragging and dropping a Sequence (Workflow ➤ Control ➤ Sequence) onto the Main screen and then putting a Message Box (System ➤ Dialog ➤ Message Box) on it. You will then enter the following in Message: "Hello, World!" But do not put in the quotes.

When you press the Start button or F5, you will get an error and the bot will stop. But what if you do not want this happen? Suppose you instead want the bot to continue but only provide you with a warning?

This is where Try/Catch comes in. Drag and drop this onto the Main screen (Workflow ➤ Error Handling ➤ Try Catch). You will then right click the top of the Log Message box and select Cut. Then go to Try and paste it inside.

You will now have this (as seen in Figure 7-17):

Figure 7-17. *This shows the use of the Try/Catch activity*

You will then click "Add new catch" and then select System.Exception, which is a generic option. Next, click to the right side and drag and drop a Log Message inside. Select Warn for the Log level and enter this for the Message: "An error has been detected." And make sure to include the quotes.

When you launch this bot, the first Log Message will be triggered (the "try"). But since it has an error, the code will execute the second Log Message (the "catch"). In other words, by doing this, the program will continue to run – but you will still be able to track the potential problems.

Common UiPath Functions

We have looked at the different types of workflows and programming structures for UiPath. But as you create bots, you will encounter certain types of scenarios like logins and screen scraping. These are at the core of RPA functionality.

So let's start with a login. Actually, UiPath has provided a test environment – called Acme System 1 – to help with this, which is at `https://acme-test.uipath.com/account/login`. It essentially emulates a company, such as with accounts for vendors, checks, and invoices. For the most part, you can fairly easily test the numerous functions of UiPath.

To use Acme System 1, press Register and enter your e-mail and password. Next, we will create a bot to login. To do this, you must have the Microsoft Edge extension installed, which you can access from the Tools menu item on the start screen of the UiPath Studio. You will then press Launch and Turn On after it is downloaded.

Go to create a bot by dragging and dropping a Sequence on the Main screen (Workflow ➤ Control ➤ Sequence). Then go to the Record at the top and select "Web." As the name implies, this will record your actions on a web site, which will be Acme System 1. After you are done, press to stop the Record and your Sequences will have the workflows for the login. As you use UiPath – or any other RPA system – the recording of actions will be quite common.

But there also must be ways to help customize this. After all, what if someone else is logging into the system? In this case, you want to allow UiPath to make the adjustment – which is done by using Arguments. This is available at the bottom Tab on the Main screen. You will specify the name of the argument – say, the username or password – and then indicate the type and default value.

Another common function of RPA is Screen Scraping. UiPath Studio actually makes it fairly easy to do this. You will select "Screen Scraping" at the top of the screen and then go to a web page that has information. For example, it may be a table of data and UiPath will automatically detect this. It will then ask if you want to scrape all of it. It will also see if the data spans multiple pages and UiPath will handle this, such as put it in a Data Table. This makes it easy to manipulate within the bot with Excel or another application.

The UiPath Orchestrator

So far in this chapter, we have covered the UiPath Studio. But once you have finished creating a bot, you will need to manage it. This is done by using the UiPath Orchestrator.

In terms of deployment, you can either have this software on your own servers or the cloud (at cloud.uipath.com). Regardless of which option you choose, you can have separate instances of the software. For example, you could have ones for development, staging, and production, which should help simplify the bot development process. Or you could have instances for different departments, say, for human resources, finance, and marketing.

In this chapter, we will look at the cloud version. So log in and you will be taken to the portal screen. Next, click Service that is at the bottom of the screen – which will take you to the Orchestrator (see Figure 7-18):

Figure 7-18. *This shows the Orchestrator*

First go to the top right of the page and click your user icon and then select Settings. In the general tab, you can choose your time zone. If you belong to a global organization, you might want to set this for "(UTC) Coordinated Universal Time." In the other parts of the Settings section, you can integrate with your e-mail system, find the details of the deployment URLs and provide for security.

On the left side of the screen, you will have the following:

- Robots: This is where you can register your bots.

- Processes: On the left side, there is a + button, which allows for the creation of new processes. This essentially brings in the bots that were created in the UiPath Studio.

- Jobs: This shows the bots that are in operation or have been in operation.

- Schedules: You can set the times for deploying the bots.

- Assets: You can create variables for the Orchestrator (text, Boolean, integer, and credential, which is for the logins).

- Queues: This is an advanced system to help manage bots.

To see how all this works, here's an example. We'll start with a simple bot that pops up a message with "Hello, world!" (we created this earlier in the chapter).

We'll then look at how to connect the bot to the cloud. First, go to the main screen and select Machines and press +. You will get three icons: "Standard Machine," "Machine Template," and "Close." We'll use the first one (which, by the way, is the most common one to use).

Before doing this, you need to locate the ID of your computer. To do this, bring up the Windows Explorer (which shows the files on your Windows system) and right-click "This PC." You will find your machine ID at "Computer Name:" or "Device Name:" You will then enter this when you select "Standard Machine" and click "PROVISION."

You will then want to get another piece of information: the machine key. You will find this by pressing the "..." (which are vertical).

Now click "ROBOTS" at the right side of the screen, click +, and you will see three icons: "Standard Robot," "Floating Robot," and "Close." Choose the first one and select the computer name from the dropdown. Then enter "Hello" as the name for the bot and then the login credentials for your Windows machine.

UiPath will update the screen to show the robot but also indicate that it is disconnected. To make the connection, you will need to launch the UiPath Robot application, which you can find on your Windows Start menu. At the top, click the settings icon and enter the machine ID and then put `https://platform.uipath.com` for the Orchestrator URL. This will then activate the bot.

Best Practices for Bot Development

Developing a bot can be a matter of a few drag and drops. But of course, if you want to design a good one that gets results, then there are some best practices to keep in mind:

- Templates and Bot Stores: Before designing a bot, see if there is already one available. The top RPA providers have their own stores as well as libraries. So do some checking first.

- Reusable Workflows: This means you can design your bot from smaller parts. Of course, this helps to reduce the complexity. But there is another important benefit: easier updates. For example, suppose when you create a bot, you repeat the creation of a certain Sequence. You may have, say, 20 bots that use it. But then later you need to make some changes to the configuration. Well, this means you will need to go back to each of the bots and make the change. However, if it had a reusable workflow, then you would just have to change this once.

- Be the Bot! Actually, designing an effective bot takes a change in mindset. It's not about completely replicating the actions of a human. One reason is that the process may be fairly inefficient. What's more, a bot will run 24/7 (a human, of course, can put in about eight hours or so). The key is that you should think like…a bot! True, this sounds kind of tough and strange. Can humans really think like a computer program? Not necessarily. But when designing the bot, it's important to take a "blank slate" approach and not just focus on typical human approaches.

- Readability: Take the time to make sure your workflows and code are easy to understand. One part of this is to set up a standard naming convention for the activities and variables. It is also a good idea to make comments within the bot that describe the different steps.

- Think About the Business: "Bot development should involve the business teams and the IT teams working together," said Asheesh Mehra, who is the cofounder and CEO of AntWorks. "This is key since RPA development should work toward automating the identified business process. These teams should be sure to employ technology and develop solutions that can scale as needed. And they should clearly define business rules, including exceptions and when a human must get involved."

Conclusion

Even though we covered lots of ground in this chapter, we still only looked at the beginner level of UiPath! The software is definitely powerful and allows for deep usage of an IT environment. Although, with what we have looked at in this chapter, you would still be able put together some useful automations.

Next we'll look at deployment and monitoring.

Key Takeaways

- Most RPA vendors have a trial version of their software and some even have free versions (although there are limitations on the usage). Such options make it easy to get a sense of the capabilities of the software.

- A traditional RPA system will have two key parts: a studio or designer (which is where you create the bot) and the Orchestrator (where you manage the bots).

- When a team develops bots, it is a good idea to use features like version control, which tracks and documents the steps in the process.

- Even the most sophisticated RPA systems cannot have every feature or integration. This is why you will have the option to install modules.

- An RPA system will have a workflow like a sequence or flowchart. This is a top-down visualization of the bot, which makes it easier to understand. It is also recommended to give descriptive names for each of the sequences and flowcharts.

- It is important to provide comments and annotations with the features/steps of a bot. After all, this will help lessen the learning curve when new developers are brought onto the team.

- You can also log messages as a way to document the code. They will then show up on the orchestrator.

- To develop a bot, you will need to understand variables. These are containers that hold data that can be manipulated. Some of the basic data types include Boolean (this is either a yes or no), integers (numbers without decimal points), strings (they hold text information), and custom-built structures (say, to hold contact information about customers). But of course, an RPA system will allow for more sophisticated data collection, such as with internal system variables.

- The scope of a variable is where it is available to be used. This is critically important since you do not want variables that could potentially impact other parts of the bot.

- Bots use other common structures of a computer language like loops and conditionals.

- A loop is a way to repeat a set of instructions until a condition is met. Some of the main ones include For Each, Do While, and While.

- A conditional will branch to a certain part of the program based on a condition (say, if a number is greater than a particular value). Some examples include IF/THEN/ELSE and Switch.

- Often debugging takes more time than creating the initial bot. In fact, the process can be quite complicated. But RPA systems provide debugging tools. They include setting breakpoints – which stops at certain points in the code – and then allowing for you to step in over different sections. You can also embed the bot in a try/catch. This will execute the code even though an error is triggered.

- Key recording is a standard feature in an RPA system. This will capture a person's keyboard/mouse movements, which can then be embedded in a bot. There are also ways to customize the process (e.g., by asking for login credentials).

- With an RPA system, you may want to have different instances of the software. This could be based on the stage of the bot creation – development, staging, and production – or the department.

- Before creating a bot, it's recommended to see if there is already a bot available, say, with a template or from a bot store.

- If a bot has a repeated workflow, then you might want to make this reusable. This helps to reduce the development time.

CHAPTER 8

Deployment and Monitoring

Ensuring Long-Term Success

A study from PEGA – which was based on a survey from more than 500 decision-makers from a diverse set of global businesses – showed that 67% thought that RPA was more effective than they originally thought (only 8% believed it was less effective than expected).[1]

Yet there was a big problem. The respondents generally considered that ramping up RPA and maintaining the effectiveness was more difficult than expected. Among the hardest challenges, according to 50% of the respondents, was bot deployment. Consider that about 39% of the deployments were on schedule and the average time for start-to-deployment was about 18 months. And once they were in production, they tended to break down (87% of the respondents said this was the case).

Another issue was maintenance of the bots. For example, a bot initiative may last only about 1.8 years. In other words, there needs to be continued investment and focus to make RPA work. The survey also showed that the bots added more complexity to the IT infrastructure.

---

[1] www.pega.com/about/news/press-releases/survey-most-businesses-find-rpa-effective-hard-deploy-and-maintain

© Tom Taulli 2020
T. Taulli, *The Robotic Process Automation Handbook*,
https://doi.org/10.1007/978-1-4842-5729-6_8

Since RPA is about automation, it's understandable that people would think this technology runs on auto pilot. And perhaps, in the future, this will be the case as the technology learns from AI.

But as of now, it is a big mistake to just think of RPA as a launch-and-forget system.

"RPA's popularity isn't surprising given its affordability, integration ease, and effective automation of repetitive tasks," said Sanjay Srivastava, who is the chief digital officer at Genpact. "For many corporations looking to start the journey on digital, RPA becomes the obvious entry point. Despite this, many companies face challenges to get enterprise-wide benefits and indeed some of the first movers are already looking at an effective restart to reboot their first attempts."[2]

Here's something else from a Deloitte survey: Among 424 companies, about 53% had begun an RPA journey but a mere 3% had actually scaled the implementation to more than 50 bots![3]

In this chapter, we'll take a look at how to beat the odds – that is, about the strategies to deploy RPA effectively across the organization.

Testing

In 2002, Washington State changed its prison sentencing to allow for credits for good behavior. This was implemented into a software system that determined the release date.

Sounds like a good idea, right? This is true. But the software had a bug. On average, it allowed the release of prisoners 49 days earlier! In some cases, it was over 600 days.

---

[2]From the author's interview with Sanjay Srivastava, who is the chief digital officer at Genpact, on October 10, 2019.

[3]www.celerity.com/how-to-scale-rpa-the-art-of-the-possible

It was not until 2015 that an IT person realized the error and it was corrected. Yet more than 3,200 prisoners had been released earlier than was required by law.[4]

This example certainly highlights the critical importance of software testing. True, it can be a tedious and boring process – but it is absolutely critical. The same goes for bot development, which can involve some thorny issues. Even a slight change in an underlying process can make a bot go haywire.

Now the temptation is to have the developers or technical people handle testing. But this is a mistake. You want someone who has a background in quality assurance for software and has experience with test automation tools (such as Selenium, Eggplant, WebTest, Mercury QTP, or Watir).

When it comes to testing, there are several approaches, which include the following:

- Black Box Testing: This is where the internal structure and design of the bot is not known to the tester. This involves setting up certain inputs and test cases – and then seeing if the results are correct.

- White Box Testing: This involves testing the internal structure and design of the bot. In other words, the tester will analyze the source code and try to detect any issues, say, security exposures, poor processes, or convoluted paths.

- Grey Box Testing: Here you have a blend of the black and white box testing approaches.

When it comes to testing bots, there is usually both manual and automated techniques. With the former, you will have a tester try many

---

[4]www.bbc.com/news/technology-35167191

different options and record them (this often requires looking at the log files, database entries, etc.). This can be time-consuming but it is still useful. There is the benefit of creativity and imagination.

As for automated testing, you will set up scripts that have different values and then run them on the software. This is certainly faster and can test for many possibilities. But automated testing is certainly constrained.

In light of all this, there is usually a combination of both manual and automated testing.

It is also recommended that there be a detailed review of the underlying processes that the RPA system is automating. If not, a tester will likely miss lots of problems. So he or she should review the documentation like the Process Definition Document (PDD) or the solution design document (SDD). And if these materials are not particularly clear? Well, then it is critical to revamp them.

Finally, there should be analysis of the data for the test scripts (the information is usually in an Excel worksheet). Is it comprehensive? Does it reflect the actual use cases?

Note that for RPA systems, you will have debugging tools to help with the testing. And in some cases, there are third-party developers who have created test automation systems. For example, Stefan Krsmanovic has done this with UiPath, which allows for running unit tests.

Going into Production

Once the testing is complete, then it's time for deployment or production. A good approach is to start with a narrow use case. This could, for example, mean just having the bot on one desktop. After some evaluation, you can then give wider distribution within the department. Then over time, there could be consideration to go enterprise-wide, depending on the type of automation involved.

"During deployment it's important to carve out a comprehensive process map and call out the specific tasks to be automated," said Asheesh Mehra, who is the cofounder and CEO of AntWorks. "Those deploying the solution will define the roles of the bots in the process. They also should keep the affected team members in confidence, iterate the process to check the infrastructure and software, and develop a fallback plan."

The fallback plan is particularly important. Let's face it, there will likely need to be tweaking of the bot.

It's also advisable to centralize the deployment process, such as through the CoE.

Monitoring

Once a bot goes live, you need to monitor it. The good news is that an RPA's orchestration system will usually have useful dashboards for this. Yet you still may want to set up your own unique metrics and KPIs to follow.

"Monitoring an RPA implementation should include measuring against the success criteria," said Mehra. "You want to know whether your deployment met those criteria or fell short. It's also important to configure the bots to handle changes and monitor the impact of that. Considering broader efficiencies and compliance requirements should also be part of your monitoring effort. Ideally, you'll want to measure to see whether employee efficiency has increased, and you'll definitely need to analyse the compliance requirements of the bot."

Also consider putting together a business continuity plan. This is a document that sets forth the types of monitoring, the goals to achieve, and the actions to take when something goes wrong. And yes, this is something that the CoE can draft and enforce.

There should also be periodic meetings to evaluate the performance of the bots. Are things tracking? Should there be changes? Should a bot even be terminated?

What's more, if you want to use more sophisticated analytics, you can do this by parsing the log files. They can be stored in a platform like Elasticsearch and you can create visualizations for location analysis with a tool like Kibana. There is also a module for machine learning (ML).

Security

Cybersecurity is becoming a notable risk with RPA. After all, some implementations are made quickly, without much collaboration with IT. The focus is often on getting to efficiencies as fast as possible. As a result, the IT department may be viewed as an unnecessary point of friction.

But this can be a big mistake. RPA is actually quite vulnerable to a cyberattack. The main reason is that a bot can have access to applications, databases, and the network. Often these may span geographies and departments. In fact, the access privileges can be much more powerful than what a typical user has.

HR should also be involved in the process. This will help deal with the management of the enterprise IDs in order to access the systems. Let's face it, you do not want there to be unexpected access to processes in areas like payroll, health insurance, and so on!

"RPA software interacts directly with critical business systems and applications, which can introduce significant risks when bots automate and perform routine tasks," said Kevin Ross, Sr., who is a system engineer at CyberArk. "Bots don't need administrative rights to perform their tasks, but they do need privileged access to log in to ERP, CRM and other enterprise business systems to access data, copy or paste information, or move data through a process from one step to the next. Privileged access without security is a recipe for disaster.

"As demand for RPA increases among lines of business, the number of privileged account credentials hard-coded into scripts or stored insecurely grows. That significantly increases the associated risks. With these approaches, the credentials end up being shared and reused repeatedly. Unlike the credentials used by humans, which typically must be changed regularly, those used by bots remain unchanged and unmanaged. Because of this, they're at risk from cyber criminals and other bad actors who are able to read or search scripts to gain access to the hard-coded credentials. They are also at risk from users who have administrator privileges, who can retrieve credentials stored in insecure locations."[5]

Note   According to a survey from EY, 74% of security professionals are concerned about insider cybersecurity threats.[6]

So if a hacker gains access to a bot, he or she could have access to highly sensitive data or engage in malicious activities like a denial-of-service attack. This may result in a company not being able to process invoices or payables, for example.

Given all this, it is imperative to have IT involved in the early stages of the planning of the RPA effort. There are also some best practices to consider, such as the following:

- Protection of Credentials: Make sure you securely store and manage them.

- Application Access: Be mindful of what the bot is doing. What would happen if it were in the control of a bad actor?

---

[5]From the author's interview with Kevin Ross, Sr., who is a system engineer at CyberArk, on October 5, 2019.

[6]www.ey.com/Publication/vwLUAssets/ey-how-do-you-protect-robots-from-cyber-attack/$FILE/ey-how-do-you-protect-robots-from-cyber-attack.pdf

- Governance: Sketch out a framework for security, which should cover the design of the bots and the use of data. There should also be a clear definition of roles and responsibilities.

- Audit Trail: You want an RPA platform that provides this, such as with the creation of logs. This means you will have a way to conduct investigations and assessments.

- RPA Security: Look at those RPA software systems that have high levels of security. For example, is there a third party that confirms that it has met certain requirements? Furthermore, it is not a good idea to use a free version of an RPA system in production. In most cases, the security will not be strong.

- Rotation: A way to help protect credentials is to change the access privileges continuously. Thus, if a hacker gains access, the usefulness will not last long.

CyberArk, for example, develops security software that provides this critical service. The rotation is provided with integrations for the main RPA software applications like UiPath, AntWorks, Automation Anywhere, AutomationEdge, Blue Prism, EdgeVerve, Focal Point, NICE, PEGA, Softomotive, and WorkFusion. But there are other important features like the creation of an audit trail and the ability to match a company's compliance policy with the underlying bots. The software also works across on-premise, cloud, and hybrid IT environments.

Despite the risks from RPA platforms, the technology ironically may help to promote security as well! We already learned about this in Chapter 1, where one of the benefits of the technology is that the tasks may have no human intervention, which minimizes fraud (of course, this is so long as there is security with the credentials). But RPA can also be effective in being agile in taking proactive actions. That is, the technology can be set up to instantly install security patches.

When it comes to security, there are definitely many approaches and strategies. With a financial services company, like Voya, the requirements are quite strict (and for good reason, as customer trust is absolutely essential for the company). "We apply the same security mechanisms for RPA that we use for any enterprise application or system," said Jeff Machols, who is the VP of the Continuous Improvement Center for Voya Financial. "We have code peer reviews and change of control. We also have separation of capabilities with the bots. We do this even though it means having to buy more licenses. Also, Voya does not have citizen developers who can essentially build whatever they want. We want to maintain control and make sure governance is being complied with."

Regardless of the overall requirements and regulatory demand, it's important to keep security as top-of-mind with RPA.

Scaling

Scaling is the holy grail of RPA. It's when automation has truly become transformative. But as we've learned at the start of this chapter, scale has proven incredibly difficult.

Why so? Well, there are a myriad of factors, such as the following:

- Planning: This is absolutely required for scaling RPA. You need to have detailed objectives and documentation (actually, it is important to emphasize this at every step in the process). If not, there will usually be chaos and confusion when there are many bots within an organization. A common way to deal with this is to retain consultants and even use process mining software. In some cases, RPA software will have its own process discovery applications. For more information on planning, check out Chapter 4 and then go to Chapter 12 to learn about process mining.

- ROI: If there is too much focus on this, then it can be difficult to scale RPA. The reason for this? Let's take an example: Suppose that a company only takes on automation of those processes where the ROI is 100% over a two-year period. This may intuitively sound like a good approach. But there is a nagging issue: What if there are only a few processes that meet this criteria and they impact a small part of the organization? In this case, when looking at a holistic view of things, there will really be little impact from the RPA effort. Rather, if the average ROI is, say, 20%, but it encompasses many parts of a company, then there could be a major impact on productivity and efficiency. This is known as the "portfolio effect." But for it to work, there needs to be a strong emphasis on change management (we covered this in Chapter 4).

- CoE: This is absolutely essential for scaling. There needs to be some discipline, governance, and centralization of the RPA implementation and monitoring. The CoE should also draft the SOP (standard operating procedures) and collect any of the learnings and best practices. There also needs to be ongoing training and education.

- IT: It's common not to involve IT in the early phases. And this is understandable as RPA technology is relatively easy to use. What's more, IT can bring friction to the process. But if you want to scale the system, then you need to have this group involved. They can be instrumental in helping with building the right technology foundation – such as with the cloud or sophisticated approaches like Kubernetes – as well as assist development techniques, dealing with updates and patches (which, by the way, is quite common with RPA), integration, and credentials management.

OK, other than have a CoE, IT involvement, and strong planning, what else should be done to scale? Granted, there is still much to be done. Scaling is a full-time effort. It's something that should be a strategic priority.

Because of this, there needs to be buy-in from the executives of the organization. This will help provide the resources and momentum for the scaling. The executives also need to be on the same page with the vision for RPA.

Something else: The CoE cannot do everything. Rather, its role is to provide leadership and guidance. This means there must be enablement of all the people who are involved with the RPA system. To do this, there needs to be guidance, training, technical support, and other assistance.

Finally, as you scale the RPA system, there will be challenges for global organizations. There will need to be attention paid to each country's own unique cultural norms as well as compliance and regulatory requirements, infrastructure readiness/reliability, privacy, data retention policies, work rules, and so on. This process will take time but will allow for wide-scale adoption – providing for maximum impact of the RPA technology.

Conclusion

We have looked at how many of the key factors of success come together, such as the need for planning and creating a CoE. Yes, it is a lot of work and can seem daunting. But when done in a step-by-step fashion, the RPA project should demonstrate considerable success.

In the next chapter, we'll take a look at data strategies.

Key Takeaways

- While RPA has proven to be quite effective, it has also been shown to be difficult to scale. Some of the reasons for this include lack of a CoE to manage the process, little planning, no meaningful involvement of IT, and too much focus on ROI.

- There should be in-depth testing of a bot before it goes into production. While RPA vendors provide debugging tools, you should probably go beyond this. This could involve using different scripts of automated testing and also having manual testing. Another approach is to have a limited pilot of the bot before it is officially rolled out.

- There are different approaches for software testing: black box (the tester does not know the internals of the bot but instead looks at the input and output to see if there are any anomalies), white box (this involves analyzing the source code), and grey box (this is a combination of black and white box testing).

- When testing, there should be a detailed review of the underlying processes, such as the Process Definition Document (PDD) or the Solution Design Document (SDD).

- Bots can easily break, which often happens because of a change in processes or updates to an enterprise infrastructure or security systems. This is why it is important to have ongoing monitoring, such as from the CoE. Keep in mind that RPA systems have dashboards for monitoring. But you could even use other tools, like Elasticsearch and Kibana, to analyze the log files, which should provide much insight.

- RPA systems are particularly vulnerable to cybersecurity threats. This is because the processes may require logins to systems that have confidential user data, such as payroll. Because of this, it is imperative to have IT involved to make sure there are security safeguards in place. Some recommendations include: protect credentials (say, with security software); allow for an audit trail; and do not use free versions of RPA software in production.

CHAPTER 9

Data Preparation

Being Ready for AI and Process Mining

Because of the power of data and algorithms, Google has created one of the most valuable companies in the world, with a market value of $900 billion. Not bad for a company that is only 20 years old.

But data can be fraught with issues. In November 2019, the Wall Street Journal published an article with the title of "Google's 'Project Nightingale' Gathers Personal Health Data on Millions of Americans."[1] This stirred up lots of controversy and the US government quickly launched an investigation.

By partnering with the Ascension chain of 2,600 hospitals, Google was able to amass data – such as on diagnoses and hospitalization records – of patients across 21 states. The goal was to find ways to improve treatments but also to lessen administrative costs.

But this begged many questions: Why wasn't this project disclosed? Were there potential privacy violations? And could Google really be trusted with this type of information?

This debate will certainly be contentious. It also is another indication that data is the "new oil."

Granted, most companies will not run into such issues or be the target of public ire. Yet the use of data and algorithms will continue to gain adoption and pose challenges and risks.

---

[1] www.wsj.com/articles/google-s-secret-project-nightingale-gathers-personal-health-data-on-millions-of-americans-11573496790

© Tom Taulli 2020
T. Taulli, *The Robotic Process Automation Handbook*,
https://doi.org/10.1007/978-1-4842-5729-6_9

Then how does this relate to RPA? Well, for a traditional implementation, there is often not a need for data analysis. But as companies expand on their efforts, this will definitely change. Keep in mind that more RPA platforms are adding AI capabilities and other sophisticated analytics. There is also the emergence of process mining, which involves crunching large amounts of event data (we will cover this in Chapter 12). Finally, RPA both consumes data and inherently creates data (in the form of detailed process recordings, system states, data access records, etc.) and requires retention of large amounts of data for compliance and audit purposes as well as performance management.

In other words, it's a good idea to have an understanding of data as well as to have an overall strategy.

Types of Data

In Chapter 2, we discussed two important types of data: structured and unstructured. When it comes to analytics and AI, the latter is generally the most important. It's this kind of data that has higher volumes, which allows for more sophisticated models. Consider that about 80% or so of data for an AI project will come from unstructured data.

This presents a tough challenge, though: You will have to find ways to essentially structure it. This is often time-consuming and prone to errors. Although, as AI becomes more important and has received much more venture funding, there are emerging new platforms to help with the process.

As for the other types of data, they include the following:

- Meta Data: This is data about data. Essentially, it is descriptive. For example, with a music file, there is metadata about the size, length, date of upload, comments, genre, and so on. All in all, this type of data can be extremely important in AI models.

- Semi-Structured Data: As the name implies, this is a blend of structured and unstructured data. Often semi-structured data has some type of internal tags for categorization. This could include such things as XML (extensible markup language), which is based on numerous rules to identify elements of a document, and JSON (JavaScript Object Notation), which is a way to transfer information on the Web through APIs. However, semi-structured data represents a small portion, say, 5% to 10%.

- Time-Series Data: This is data that is ordered by time. In terms of AI, this information can prove quite useful for predictions. But time-series data is also crucial for applications like autonomous driving and tracking the "customer journey".

Big Data

Big Data is certainly a big business. According to IDC, the global spending on this category (including business analytics) is forecasted to hit a whopping $274.3 billion by 2022.

In the report, IDC group's vice president of Analytics and Information Management, Dan Vesset, had this to say: "Digital transformation is a key driver of BDA (big data and business analytics) spending with executive-level initiatives resulting in deep assessments of current business practices and demands for better, faster, and more comprehensive access to data and related analytics and insights. Enterprises are rearchitecting to meet these demands and investing in modern technology that will enable them to innovate and remain competitive. BDA solutions are at the heart of many of these investments."[2]

---

[2]www.idc.com/getdoc.jsp?containerId=prUS44998419

So what is Big Data? Actually, the definition is somewhat elusive. Part of this is due to the fact that Big Data involves constantly evolving innovation. What's more, the category is massive, as it has multiple use cases.

Here is how SAS defines it: "Big data is a term that describes the large volume of data – both structured and unstructured – that inundates a business on a day-to-day basis. But it's not the amount of data that's important. It's what organizations do with the data that matters. Big data can be analyzed for insights that lead to better decisions and strategic business moves."[3]

As for the origins of the term Big Data, it goes back to the early 2000s. This is when we saw the emergence of cloud computing and the growth in large Internet apps. There was a need to find ways to manage the enormous data volumes, such as with open source projects like Hadoop.

Now another way to look at Big Data is through the prism of the three V's (this was developed by Gartner analyst Doug Laney). They include the following:

- Volume: Yes, there must be a massive amount. While there is no bright line for this, there is some consensus that Big Data should be in the tens of terabytes.

- Variety: Big Data has much diversity, such as with structured, semi-structured, and unstructured data. But as we've noted earlier, unstructured data is the main source.

- Velocity: This is the speed that data is received and processed. To handle this effectively, Big Data will often leverage in-memory approaches.

---

[3]www.sas.com/en_us/insights/big-data/what-is-big-data.html

Given the size and diversity of the Big Data business, it should be no surprise that more V's have been added! One is veracity, which goes to the overall accuracy of the data. This will be a key theme in this chapter, such as with data cleansing.

Then there is visualization. This involves the use of graphs to better understand the patterns in Big Data. This is a major part of BI (business intelligence) systems.

And then there is variability, which is concerned about how data changes over time. An example of this is the evolution of sentiment with social media content.

The Issues with Big Data

Now even with all the advancements in storage and software systems, Big Data is still challenging for many companies – even for those that are technologically savvy. Let's face it, the growth is only accelerating.

Based on data from Cisco, the annual global IP traffic will hit 4.8 zettabytes per year.[4] To put this into perspective, it was 1.5 zettabytes in 2017.

And what is a zettabyte? It is 1,000,000,000,000,000,000,000 bytes or a billion terabytes or a trillion gigabytes! It's enough to store 36,000 years of HD-TV video.

Thus, when it comes to parsing through the huge amounts of data, there are major complexities. As a result, it's not uncommon for the projects to have notable failures.

Another issue is talent. The fact is that it is expensive to hire data scientists, who have a solid understanding of advanced statistics and machine learning. Because of this, many companies simply are unable to capitalize on their datasets effectively. However, there are companies such

---

[4]www.cisco.com/c/en/us/solutions/collateral/service-provider/visual-networking-index-vni/white-paper-c11-741490.html

as Alteryx that are trying to solve this problem, such as by building tools that nontechnical people can use.

What's more, the underlying core technologies for Big Data have seen significant changes over the years. There are not just the mega tech providers like Amazon, Google, and Teradata but also a myriad of start-ups, such as Snowflake. There has also been tremendous evolution in open source platforms, with Apache Spark getting lots of traction.

Finally, it's important to keep in mind that you do not necessarily need huge amounts of data for good results. Even smaller sets can yield strong outcomes. The key is having an understanding of the business use case.

Note   According to research from Gartner, about 85% of Big Data projects are abandoned before they get to the pilot stages. Some of the reasons include lack of buy-in from stakeholders, dirty data, investment in the wrong IT tools, issues with data collection, and lack of a clear focus.[5]

The Data Process

There are various approaches for the data process. But there is one that has much support and has been refined over the past 20 years. It is called the CRISP-DM (cross-industry standard process for data mining) Process, which is the result of a consortium of experts, software developers, consultants, and academics. Here are the main steps, as seen in Figure 9-1:

---

[5]www.techrepublic.com/article/85-of-big-data-projects-fail-but-your-developers-can-help-yours-succeed/

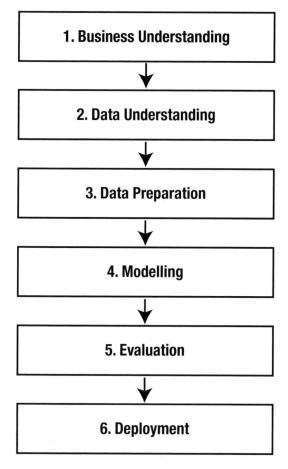

Figure 9-1. *These are the steps in the CRISP-DM Process*

Consider that these steps are not set in stone. When going through this process, you might iterate on one step several times – or go back to an earlier step.

But the key is that there is an overall plan.

OK then, before you start on the CRISP-DM Process, you need to assemble the right team. Ideally, you should have one or two data scientists – or those who have a solid technical background. The good news is that there are many courses, such as on Udemy, Udacity, and

Coursera, that can train your team on data science. Then you will have several people from the business side of your organization who can bring real-life experience to the project.

In terms of the data science team members, you will be competing against companies like Facebook or Google who can offer generous salaries and equity packages. Sounds kind of intimidating? Unfortunately, it can be. But then keep in mind that you do not need PhDs with experience in areas like ML and AI. Instead, you want those who have a good understanding of the fundamentals of putting together data projects, such as with using tools like TensorFlow.

Now then let's take a deeper look at the CRISP-DM Process:

Step #1 – Business Understanding

There must be a clear statement of the problem to be solved. This could be how a change in the price can lead to an improvement in sales or how better engagement can mean reduced churn. Boiling things down, a data project should have a hypothesis to test (you do not want to complicate things with multiple factors to measure).

Step #2 – Data Understanding

Here you will identify the relevant sources of data for the project. There are three categories to consider:

- In-House Data: This can come from many sources, such as the web site, mobile apps, and IoT sensors. In-house data is not only free but has the advantage of being customized to your business. Yet there are still some nagging issues, such as the challenge of formatting the data (especially if it is unstructured) and not having enough to perform useful analytics.

- Open Source Data: This is publicly available data that is often free or has a low cost. Common examples include datasets from the government (which, by the way, can be extremely useful), nonprofits, and universities. Open source data is usually formatted and comprehensive. However, there could be problems with bias (we'll discuss this topic later in the chapter).

- Third-Party Data: This is from a commercial vendor. It's true that the fees can be far from cheap. The irony is that the data may not even be particularly useful or accurate! So before purchasing third-party data, it is important to do some due diligence.

Note   Based on the AI engagements of Teradata, about 70% of data is from in-house sources, 20% from open sources, and the remaining from commercial vendors.[6]

When evaluating data, here are some questions to ask:

- Is the data complete?
- What might be missing?
- Where did the data come from?
- What were the collection points?
- Who touched the data and processed it?
- What have been the changes in the data?
- What are the quality issues?

    Step #3 – Data Preparation

---

[6]This is from the author's interview with Teradata's Atif Kureishy in February 2019.

After you have selected your data sources, you will then need to take steps to cleanse it. This is where having expertise with data science will be critical. Even small errors can lead to terrible results in a model.

According to Experian, bad data has had a negative impact on 88% of US companies. The research indicates that the average loss of revenues is about 12%![7]

Then what are some of the actions you can take to improve the quality of the data? Let's take a look:

- De-duplication: Duplication is a common problem with datasets. For example, a customer may change addresses or even his or her name. This could lead to data that is extraneous or misleading when scaled across a large dataset.

- Consistency: Some categories of data may not be clearly defined. For example, the word "profit" may have different meanings. So when it comes to the dataset, make sure the definitions for each parameter are understandable.

- Outliers: This is when some data is way beyond the range of the overall dataset. True, this may ultimately be fine. After all, there are exceptional cases (say, a person with a high IQ). But outliers could point to problems as well. Might the data be incorrect – such as from a bad input?

- Validation Rules: In some cases, the data will be clearly false. This would be, for instance, if a person's height is a negative number. To deal with this, you can establish certain rules and then the data can be flagged for the deviations.

---

[7]www.smartdatacollective.com/lessons-can-learn-bad-data-mistakes-made-throughout-history/

- Binning: When analyzing data, does it really matter if a person is, say, 25 or 27 years old? Probably not. Instead, you will probably want to look at broader categories (ages 30 to 40 and so on).

- Global Data: This can certainly present some tough problems, such as with the differences in cultures and standards. For example, while the United States specifies dates according to a day–month–year structure, Europe has a year–month–day approach. To manage such things, you can set up conversion tables.

- Merging: A column of data may be similar to another one. This could be if you have one that expresses height in inches and another in feet. In these situations, you might select just one or merge the columns.

- Staleness: Is the data really applicable and relevant? Is it too old?

- One-Hot Encoding: With this approach, you can replace categorical data with numbers. Here's an illustration: Suppose you have a dataset with a column that has three possible values: Phone, desktop computer, and laptop. You can certainly represent each with a number. But this could pose a problem – that is, an algorithm may think that a laptop is greater than a phone. So by using one-hot encoding, you can avoid this situation. You will create three new columns: is_Phone, is_Desktop_Computer, and is_Laptop. That is, if it is a laptop, you will put 1 in the is_Laptop column and 0 in the rest.

All this can seem overwhelming. It is also a process that can take much time. But to help with this, a good approach is to analyze a sample of the data first and search for some of the potential issues.

Next, there are a myriad of data software tools that can help out, such as from Oracle, SAS, and IBM. There are also open source projects like OpenRefine, plyr, and reshape2.

Steps #4 and #5 – Modelling and Evaluation

It's now time to look at how to create models with the data. This is done by applying algorithms to it, which are mathematical systems that can involve many steps. The temptation is actually to use the more sophisticated ones! Yet this could prove to be a mistake. It's actually not uncommon for simple models – say, using traditional statistical techniques like regressions – to get strong results.

Keep in mind that there are hundreds of different types of algorithms. But a data scientist can use tools to test them out – such as TensorFlow, Keras, and PyTorch – and also engage in trial and error.

Once a model is selected, then it must be trained with data. While much will have already been done with the data, there are still some things to consider. For instance, you do not want it to be sorted. Why? For an algorithm, it may seem as a pattern. Rather, it is better to randomize the data.

When training the model, there will be two datasets:

- Training Data: This will be about 70% of the complete dataset. The training data will be the information that the algorithms find the underlying patterns. Example: Suppose you are creating a model for predicting the price of a home. Your dataset will have variables like the number of rooms, amenities (like a pool), and the crime rate. By processing the training data, the algorithm will calculate the weights for each of the factors into an equation. If it's a linear regression, then it will look something like $y = m * x + b$.

- Evaluation of the Model: Creating models can
certainly be tricky. Consider that there may be issues
with overfitting and underfitting of the data, which
is when the algorithms are skewed. This is why it is
essential to evaluate the model with test data, which
accounts for 30% of the total dataset. For it to be
effective, the information needs to be representative.
So with the evaluation, you will look for discrepancies.
How accurately is the model reflecting reality?
Unfortunately, in some cases, there may not be enough
good data to get satisfactory results.

The training of the model requires much fine-tuning and tweaking.
This is just one of the reasons an experienced data scientist is so
important. While there are more automation tools for modelling – say,
from DataRobot – there is still a need for human expertise.

Step #6 – Deployment

The deployment of a model is either within an IT infrastructure or as
part of a consumer-facing app or web site. Because of the complexities and
risks, it is probably best to deploy the system on a limited basis (perhaps
with beta users). Rushing a project is usually a recipe for failure.

Note that even the world's best technology companies have failed
miserably when it comes to deployment. Here are some notable examples:

- Microsoft's Tay: This was a chatbot on Twitter that was
launched in March 2016. But unfortunately, it quickly
spewed racist and sexist messages! The problem was that
Twitter trolls were able to manipulate the core AI. As a
result, Microsoft took down Tay within about 24 hours.

- In March 2009, a shooter live-streamed on Facebook his horrific killing of 50 people in two mosques in New Zealand. It was not actually shut down until 29 minutes after. Simply put, Facebook's AI was unable to detect it. The company would write a blog, saying: "AI systems are based on 'training data,' which means you need many thousands of examples of content in order to train a system that can detect certain types of text, imagery or video. This approach has worked very well for areas such as nudity, terrorist propaganda and also graphic violence where there is a large number of examples we can use to train our systems. However, this particular video did not trigger our automatic detection systems. To achieve that we will need to provide our systems with large volumes of data of this specific kind of content, something which is difficult as these events are thankfully rare. Another challenge is to automatically discern this content from visually similar, innocuous content—for example if thousands of videos from live-streamed video games are flagged by our systems, our reviewers could miss the important real-world videos where we could alert first responders to get help on the ground."[8]

When it comes to deployment, there are often tough issues with existing systems. Usually they were not built for deploying algorithmic applications and may use different programming languages. There will also need to be a change in mindset as IT people may not be familiar with using technologies for making predictions and insights based on large amounts of data. Because of all this, it is often required for there to be rewriting and customization as well as some training.

---

[8]https://newsroom.fb.com/news/2019/03/technical-update-on-new-zealand/

Another issue is that there may not be enough attention paid to the end user, who usually does not have much of a technical background. This is why a model should have an easy-to-use interface and workflow.

You should also avoid any jargon or complex configuration. You want as little friction as possible. To this end, try to limit the options for the end user, which should help with adoption. There should also be documentation and other educational resources, such as videos. For the most part, you will need to engage in change management as well (we covered this in Chapter 6).

Finally, the deployment of models is an iterative process. You will need to have periodic monitoring to see if the results continue to be accurate.

Types of Algorithms

As noted earlier, there are many types of algorithms. But you can narrow them down into broad categories, which should provide an easier framework for using them effectively.

Here's a look:

#1 – Supervised Learning

This is where algorithms use data that is labeled. They are the most common types and also generally require large amounts of data to be effective. But unfortunately, it can be tough to come up with this type of information. In fact, one of the biggest stumbling blocks for AI has been the time-consuming nature of creating labeled datasets.

But there are some strategies for dealing with this. One is from Fei-Fei Li, who is a PhD from Caltech who would go on to specialize in creating AI models. She at first had her students come up with the datasets but this did not work out. Although, one of them suggested she try crowdsourcing, similar to Amazon.com's Mechanical Turk. She tried it out and was able to amass a database of 3.2 million labeled images (for over 5,200 categories). It turned out to be a game changer in the AI world and would be the source of contests to test new models. In 2012, researchers from the University of Toronto – Geoffrey Hinton, Ilya Sutskever, and Alex Krizhevsky – applied their deep learning algorithms against ImageNet and the results were startling as there was a material improvement in image recognition. If anything, this would become a pivotal moment in the acceleration of the AI movement.

Next, another way to create labeled data is to actually use sophisticated algorithms! That is, they should be able to discern complex patterns and fill in the gaps.

#2 – Unsupervised Learning

This is where algorithms are applied against data that is unlabeled. While there is much of this data available – and it is quite useful for models – it is difficult to use. Unsupervised learning is still in the nascent stages.

However, with the advances in deep learning, things are definitely improving. Algorithms can essentially find clustering in the data, which may indicate important patterns. This approach has been effective in areas like sentiment analysis – where there are patterns across social media – and recommendation engines (say, to find movies you might be interested in).

When it comes to the next generation of AI, unsupervised learning will certainly be key. In a paper in Nature by Yann LeCun, Geoffrey Hinton, and Yoshua Bengio, the authors note: "We expect unsupervised learning to become far more important in the longer term. Human and animal learning is largely unsupervised: we discover the structure of the world by observing it, not by being told the name of every object."[9]

#3 – Reinforcement Learning

---

[9]https://towardsdatascience.com/simple-explanation-of-semi-supervised-learning-and-pseudo-labeling-c2218e8c769b

Reinforcement learning is kind of like the concept of "learning my osmosis." So when you are new to something, you will experiment, trying different approaches. Over time, you will get a sense for what works and what to avoid.

Of course, regarding reinforcement learning, the underlying mathematics is extremely complex. But the overall idea is straightforward: it is about learning by experiencing rewards and punishments for actions.

Interestingly enough, some of the major advances in reinforcement learning have come in gaming. Just look at DeepMind, which is owned by Google. The company has several hundred AI researchers that have tested their theories on well-known games, such as Go. In 2015, DeepMind created AlphaGo that was able to beat the European champion (five games to zero). This was the first time this had happened – and it was a shocker. The conventional wisdom was that a computer simply would not have enough processing power to defeat a world champion. By the way, a couple years later, AlphaGo would win a three-game match against Ke Jie, who was ranked No. 1 in the world.

So why are games good for testing? They are constrained, such as with a board and a set of rules. In a way, it's a small universe.

The presumption is that a game could be the basis for understanding broader ideas of intelligence. Actually, companies like DeepMind have extended their models to real-life applications, such as in reducing energy usage and the detection of diseases.

#4 – Semi-Supervised Learning

This is a blend of supervised and unsupervised learning. This is where you may have a small amount of unlabeled data. Then with the rest you will try to label it using algorithms (known as pseudo-labeling).

A case study of this comes from Facebook. At its F8 developers conference in 2018, the company demonstrated semi-supervised learning by leveraging its massive Instagram hashtag data.[10] Basically, it served as a form of labeling, such as a description of a photo. But then again, there were some hashtags that really were not very helpful, like #tbt (which stands for "throwback Thursday"). Despite this, Facebook's researchers were able to build a database of 3.5 billion photos that had an accuracy rate of 85.4%.

Looking to the future, the company believes that this innovative approach could be helpful in improving the rankings of the newsfeed, the detection of objectionable content, and the creation of captions for those who are visually impaired.

The Perils of the Moonshot

IBM's Watson is perhaps the most well-known AI system. In 2011, it was able to beat two of the all-time *Jeopardy!* champions.

While this generated tons of buzz, IBM wanted to extend the capabilities of Watson, such as to the healthcare industry. In 2013, the company announced a strategic alliance with the University of Texas MD Anderson Cancer Center to leverage the power of AI to conquer diseases like leukemia.

In the press release for the deal, the general manager of IBM Watson Solutions, Manoj Saxena, said: "IBM Watson represents a new era of computing, in which data no longer needs to be a challenge, but rather, a catalyst to more efficiently deploy new advances into patient care.

---

[10]https://engineering.fb.com/ml-applications/advancing-state-of-the-art-image-recognition-with-deep-learning-on-hashtags/

"By helping researchers and physicians understand the meaning behind each other's data, we can empower researchers with evidence to advance novel discoveries, while helping enable physicians to make the best treatment choices or place patients in the right clinical trials."[11]

While this was certainly a bold endeavor, the results came up very short. Part of the problem was the paucity of data with rare cancers. There were also issues with the updating of the system. The bottom line: Watson would sometimes produce inaccurate information (this was according to a report in the Wall Street Journal).[12]

The University of Texas would spend about $62 million on the program but would abandon it in 2017.[13]

This is a cautionary tale of the very real limits of AI. The fact is that the technology remains fairly narrow in terms of the use cases – and far from intelligent. Yet this is not to say that AI is not worth the effort. It definitely is. But the key is to be realistic about its capabilities.

Bias

Bias in data is when the dataset does not accurately reflect the population. This could easily lead to unfair outcomes, such as discrimination based on gender, race, or sexual orientation.

---

[11]www.mdanderson.org/newsroom/md-anderson--ibm-watson-work-together-to-fight-cancer.h00-158833590.html

[12]www.wsj.com/articles/ibm-bet-billions-that-watson-could-improve-cancer-treatment-it-hasnt-worked-1533961147

[13]www.utsystem.edu/sites/default/files/documents/UT%20System%20Administration%20Special%20Review%20of%20Procurement%20Procedures%20Related%20to%20UTMDACC%20Oncology%20Expert%20Advisor%20Project/ut-system-administration-special-review-procurement-procedures-related-utmdacc-oncology-expert-advis.pdf

Bias is actually often called "the silent killer of data." And why so? The reason is that bias is often unintentional. Data scientists do not set out to create faulty models. However, their work will certainly reflect some of their backgrounds. So if the researchers are mostly white males who come from the middle class or wealthy families, then some of their tendencies will show up in their work.

Because of this, more companies are taking actions to deal with bias. "For our AI teams, we make sure we have a diverse group in terms of geographies, race, and gender," said Seth Siegel, who is the partner of artificial intelligence and automation at Infosys Consulting. "We believe that this makes our projects much more robust and effective."[14]

Or look at Disney. The company has formed an alliance with the Geena Davis Institute on Gender in Media. Her organization has developed an AI tool called GD-IQ: Spellcheck for Bias, which is based on the research from the University of Southern California Viterbi School of Engineering. The app will analyze all the content and detect instances where there are, say, too many lead male roles or there is a lack of diversity.[15]

Such efforts are encouraging. Having a tool can also help systemize the process and also provide some objectivity.

Yes, the topic of bias can be very divisive and controversial. Besides, few software developers take courses on this topic while they are getting their engineering degrees!

This is why it's important for companies to take a leadership role. Some will even set up ethics boards. But even if this is not practical, it is still a good idea to set forth some core principles, provide training, and have periodic discussions, especially when developing new models.

---

[14]From the author's interview with Seth Siegel, who is the partner, artificial intelligence and automation at Infosys Consulting, on December 5, 2019.

[15]www.hollywoodreporter.com/news/geena-davis-unveils-partnership-disney-spellcheck-scripts-gender-bias-1245347

Conclusion

Again, RPA does not necessarily need sophisticated data approaches, at least for the basic functions. But most companies want to go beyond this. RPA can be a pathway to AI and digital transformation. So it is essential to put a data strategy together.

As for the next chapter, we will look at the landscape for the RPA vendors.

Key Takeaways

- Having a data strategy is not a prerequisite for RPA (at least for when using the traditional capabilities). But if you want to move into AI or process mining, then you will need to get serious about data. Interestingly enough, RPA is often referred to as the "gateway drug" for AI.

- Besides structured and unstructured data, there are other flavors like metadata (data about data), semi-structured data (this is a mash-up of structured and unstructured data, which often includes internal tags for categorization), and time-series data.

- Big Data emerged in the early 2000s, as new technologies like cloud computing started to take off. The definition can be fuzzy. But Big Data is often explained by using the 3 V's: volume (there needs to be large amounts of data, such tens of terabytes), variety (there is diversity, say, with structured, unstructured, and semi-structured data), and velocity (this is the speed of the data).

- Big Data projects can be extremely complex and failure is fairly common. This is why there needs to be much planning and a focus on the business case.

- One approach for planning is to use the CRISP-DM Process, which has seven steps: business understanding (what is the problem to be solved?), data understanding (this involves the selection of data sources, such as in-house, private, or publicly available), data preparation (this is where you take steps to clean up the data), modelling (here you determine the right algorithms), evaluation (you will use systems to test the model), and deployment.

- Training data is the dataset you use to create a model. You will then evaluate it by using a separate dataset, which is called test data.

- Some of the general types of algorithms include supervised learning (where the data is labeled), unsupervised learning (sophisticated techniques to find patterns in unlabeled data), reinforcement learning (this involves a reward–punishment system), and semi-supervised learning (this is where there is some labeled data).

CHAPTER 10

RPA Vendors

A Look at the Major Players

They are called the "Big 3 of RPA": UiPath, Automation Anywhere, and Blue Prism. These companies are the pioneers of the industry, having amassed large customer bases. They also have a full-suite of software. Of course, they have the benefit of substantial amounts of capital.

When reading about RPA, it seems like the Big 3 are the only players that matter! As a result, this makes it tough for the many other operators to rise above the noise and get the attention of potential customers. The RPA market is showing signs of a "winners take most" dynamic.

Now this may not necessarily last. With the funding environment continuing to be robust, there is much potential for upstart RPA companies to get momentum. After all, there is likely to be considerable impact from technologies like AI and perhaps open source approaches.

In this chapter, we will look at the Big 3 but we will also cover various other providers. Keep in mind that there is quite a bit of diversity in the industry.

UiPath

Daniel Dines grew up in Romania during the 1970s and 1980s, which was under the authoritarian control of Nicolae Ceausescu. There was not much opportunity in the country. But Daniel did discover something that would change his life in a big way: the computer. Well, he actually borrowed it

© Tom Taulli 2020
T. Taulli, *The Robotic Process Automation Handbook*,
https://doi.org/10.1007/978-1-4842-5729-6_10

from a friend and learned C++ on his own. Because of his mathematical skills, he excelled at coding and was able to land a job at Microsoft. It turned out to be another transformative experience as he learned how to develop enterprise software.

In 2005, Dines decided to launch his own company, which provided outsourced technical services. But it was tough to get traction because of the competition, which resulted in steep discounting of fees. He pivoted by selling software that helped with integration.[1] Yet this also proved tough to get much momentum.

Interestingly enough, one of Dines' customers reached out to him and said that they used his software to help automate human interactions for the office. So he got on a plane and took a trip to India where the company was based and spent three months there learning everything he possibly could. What Dines got was a front-row view of the next-generation approach to RPA, which was much more than just about screen scraping and recording keystrokes.

When Dines came back to Romania, he was laser focused on this new opportunity. It would certainly prove to be yet another transformation – and by far, the biggest in his tech career.

Along the way, Dines used many of the lessons learned from Microsoft. For example, he wanted to create a similar vision to "a computer on every desk" to a "bot for every person."

Fast forward to today: The UiPath platform is full-featured for enterprises of any size, allowing for the creation of attended and unattended bots. There is also an easy-to-use designer (UiPath Studio) and system for the control, management, and security (UiPath Orchestrator).

---

[1] www.forbes.com/sites/alexkonrad/2019/09/11/from-communism-to-coding-how--daniel-dines-of-7-billion-uipath-became-the-first-bot-billionaire/

To make this all happen, UiPath has taken an aggressive approach to product development. Note that the company operates on a bimonthly release cycle.

Some of the most recent additions to the platform include:

- UiPath StudioX: This is for nontechnical employees who can create sophisticated bots without the need for using developer resources or coding.

- UiPath Apps: With this, you can provide for human interaction across any point in a process for an unattended bot. UiPath calls this "human in the loop."

- UiPath Insights: This is a set of analytics to help measure progress with an RPA implementation.

- UiPath Connect Enterprise: This uses gamification and crowdsourcing to collect ideas for automation and other innovations.

Besides this, there are other general features to note about the UiPath platform, such as:

- Security: UiPath's 18.4.4 and successive versions have been certified as meeting the highest security requirements from Veracode. Consider that from the early days the company has built the platform with defense-grade security and auditing, such as providing for encryption and role-based access control. In fact, there are 60+ government agencies that have confirmed the security of UiPath.[2]

---

[2]www.uipath.com/newsroom/uipath-rpa-security-certified-at-highest-level-of-veracode-uipath

- Integrations: They are extensive, spanning categories like BPM, image analysis, language, IoT (Internet-of-Things), machine learning, and security. The company has also been aggressive in forming partnerships, such as with Salesforce.com's AppExchange. This is for the UiPath Connector that helps to streamline the workflows – with the deployment of bots – that are native within the Salesforce.com platform.[3]

- UiPath Academy: This is the learning center, which provides 15 free courses as well as biweekly one-hour Q&A webinars. The academy has more than 360,000 students enrolled. Although, within the next five years, the goal is to train 750,000 in America. To achieve this goal, UiPath has formed partnerships with colleges and universities, such as the California State University, Fullerton, and the College of William & Mary (which involves investing more than $4 million to give "one robot to every student").

"In our RPA implementation, we initially evaluated six vendors," said Jeff Machols, who is the VP of the Continuous Improvement Center for Voya Financial. "We chose UiPath because of the breadth of the product and balance of the attended and unattended capabilities. We were also impressed with the vision, product roadmap and the culture of the company, all of which set it apart."[4]

---

[3]www.uipath.com/newsroom/uipath-announces-the-uipath-connector-on-salesforce-appexchange

[4]From the author's interview with Jeff Machols, who is the VP of the Continuous Improvement Center for Voya Financial, on November 1, 2019.

UiPath definitely stacks up quite well in terms of the evaluation from top research firms and other third-party organizations. Here is just a sample:

- In Gartner's Magic Quadrant report for RPA, UiPath was ranked as No. 1 in the industry. It was also recognized as the customers' choice, which was based on verified end users.[5]

- UiPath was the leader in the Forrester Wave report, which included 15 vendors. The software received the highest scores for "Market Presence" and "Strategy." There were also top scores for RPA functions like bot development, core UI, and desktop functions. According to Forrester: "UiPath sits in the cockpit of the RPA rocket ship. UiPath's financial backing, and savvy development and marketing, make it hard to bet against... References report that UiPath will go the extra mile to meet a client's need. They also applaud the low cost of getting started, well-organized partner channel, overall product stability and strong security."[6]

- UiPath took the No. 3 ranking on the Forbes 2019 Cloud 100 list. This was based on the feedback and analysis of Bessemer Venture Partners and Salesforce Ventures.[7]

While RPA is powerful, it is limited in its scope. But UiPath has been focused on going beyond typical functions. To this end, the company has pulled off several important acquisitions.

---

[5]www.uipath.com/newsroom/uipath-named-a-july-2019-gartner-peer-insights-customers-choice-vendor

[6]www.uipath.com/newsroom/rpa-leader-forrester-wave-2019

[7]www.uipath.com/newsroom/uipath-earns-no.-3-spot-on-the-forbes-2019-cloud-100-list

One was for ProcessGold, which is a leader in process mining. This is about how to scan, analyze, and improve upon business processes by using software technologies (we discuss this more in Chapter 12). ProcessGold is based in Eindhoven, which is essentially the Silicon Valley of the Netherlands and the place where process mining was born. Over the years, the company has introduced various innovations, such as the TRACY algorithm that allows for process graph visualizations.[8]

Then there was UiPath's acquisition for StepShot, which is a top player in the process documentation software industry.[9] It initially started as an open source project in 2009 and then morphed into a company six years later (based in Tallinn, Estonia).

With the StepShot app, it is much easier to develop guides and tutorials for RPA implementations. What may take several hours can instead be done in minutes. How? StepShot tracks steps and describes the processes – which are all then automatically translated into a professional document.

So yes, UiPath has integrated these two technologies with the Explorer system, allowing for better streamlining of the frontline and backline operations of an organization.

While broadening of the scope of RPA is definitely important for the company's growth, it does have its issues. Hey, is the name of the category really representative anymore? Or will there be a need to come up with an alternative to RPA?

"This is something we think about," said Diego Lomato, who is the VP of product marketing at UiPath. "We believe that the best approach is to evolve the term RPA. We view it as end-to-end automation."[10]

---

[8]www.uipath.com/newsroom/uipath-acquires-process-gold-unparalleled-process-understanding

[9]www.uipath.com/newsroom/uipath-acquires-stepshot-adds-process-documentation-to-market-leading-rpa-platform

[10]From the author's interview with Diego Lomato, who is the VP of product marketing at UiPath, on November 5, 2019.

This certainly makes sense. Note that this has been the case with other enterprise categories. Let's face it, CRM (customer relationship management) is much different than it was when Salesforce.com was founded 20 years ago.

Despite this, there are already attempts to completely redefine RPA. Some examples include terms like DPA (digital process automation) and IPA (intelligent process automation). But interestingly enough, such attempts may ultimately be a way for smaller players to get more attention!

But for the most part – given all the media attention and money invested – RPA is likely to be the top-of-mind way for describing software-based automation in the workplace.

OK then, what about UiPath's AI strategy? Well, according to Lomato: "The second most asked question we get from customers is, 'What is AI and what does it mean for RPA?' And as for the most common question, it is actually, 'What is RPA?'"

For UiPath, AI is not about a generalized form of the technology, such as to create a bot that can understand commands like Siri. Rather, the company is focused on how to use AI for specific ways to better accomplish tasks, say, with reading and interpreting documents and moving them to the right places. "We use very pragmatic models," said Lomato.

The centerpiece of the company's AI strategy is the UiPath AI Fabric, which includes over 100 prebuilt modules called AI Skills that you can drag and drop into your workflow. Consider that – when it comes to an AI implementation – deployment is often the most challenging part of the process. But with UiPath, it is fairly simple. It's also possible to transfer a company's own AI models into a bot.

"By being within UiPath," said Craig Dewar, who is the director of platform marketing at UiPath, "you can easily do model versioning. This is extremely helpful in validating the outcomes."[11]

---

[11]From the author's interview with Craig Dewar, who is the director of platform marketing at UiPath, on November 5, 2019.

There is something else that is important: UiPath's scale means it has access to enormous amounts of data across many use cases and industries (there are over 400,000 users). In other words, this will be a key benefit for the company's AI efforts.

Automation Anywhere

In 2003, Ankur Kothari, Mihir Shukla, Neeti Mehta, and Rushabh Parmani founded Tethys Solutions, LLC, which was based in San Jose, California. They all had strong backgrounds in Silicon Valley, having worked at companies like Intuit, Netscape, Siebel, and Infoseek.

The initial product was launched on a Sunday, and within one hour the company snagged its first customer. At first, Automation Anywhere focused on the small- and medium-sized business market, which saw lots of growth (getting to over 25,000 customers).

But Shukla – who was the CEO – realized that the real opportunity was to target large enterprises, which were looking for solutions that could transform their organizations with more efficiencies and better customer responsiveness.

The company would eventually change its name to Automation Anywhere and the vision was as follows:

"Take the robot out of the human. To liberate employees from mundane, repetitive tasks, allowing them more time to use their intellect and creativity to solve higher order business challenges. We see a world where every employee will work side by side with digital workers, making them exponentially more productive and far more fulfilled."[12]

This vision is very personal for Shukla. While growing up in India, he had few opportunities for career advancement. But he did have a chance to visit the United States where he learned about computers. "When I started

---

[12]www.automationanywhere.com/company/about-us

Automation Anywhere, I wanted to find a way to allow millions of people to benefit from technology," he said. "I wanted to pass on what benefited me. This would change lives. For me RPA could do this as it is a next-generation platform to do things and get things done."

Automation Anywhere now has more than 3,100 customer entities and 1,800 enterprises that use its platform. The technology has been effective for diverse industries like financial services, insurance, healthcare, technology, manufacturing, and telecom. To get a sense of the success of the platform, consider that Automation Anywhere customers include 85 percent of the world's top technology companies, more than 85% of the world's top banks and financial services companies, and 80% of the world's top telcos.

With 20 locations across the world, Automation Anywhere is a true global operation. The company also has more than 1.5 million bots in operation, which is more than any other RPA company.

No doubt, a key to success has been an obsession with meeting the needs of its customers, which has helped to drive innovation. Automation Anywhere has an NPS (Net Promoter Score) of +67, which is three times higher than the average for a B2B software company.

"Our goal is to get to 73," said Shukla. "The reason is that this is the level where Apple is at."

It's important to note that the company has developed its A+ Customer Success Program that provides 24/7 service. With this, every customer gets at least one support representative who is not compensated based on commissions but instead on the success of the customer.[13]

As a long-time player in the RPA space, Automation Anywhere has had many iterations of its platform (11 versions). They were all downloadable software applications.

---

[13]www.automationanywhere.com/company/press-room/automation-anywhere-posts-net-promoter-score-of-67-exceeding-technology-industry-benchmark

But in October 2019, the company made a major announcement: the launch of Automation Anywhere Enterprise A2019 (see Figure 10-1). It represented the RPA industry's first purely web-based, cloud-native platform. The company refers to it as RPA-as-a-Service.

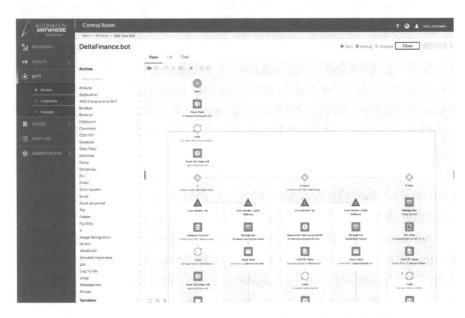

Figure 10-1. *Here's the Automation Anywhere cloud platform*

Yet this is not to imply that the company is abandoning its traditional on-premise versions. Automation Anywhere has noted that it will maintain them for perpetuity. What's more, the cloud and on-premise versions of Automation Anywhere Enterprise A2019 have been built to work together seamlessly.

Now the cloud version of the software – which works for public, private, and hybrid clouds – does seem somewhat late. Cloud computing has been around for quite some time and has transformed enterprise software categories like CRM (customer relationship management) and ERP (enterprise resource planning). But such technologies are for discrete purposes and work mostly with structured data.

RPA, on the other hand, must work deeply with the internals of desktops and applications. As a result, it is difficult to develop a cloud version.

As for Automation Anywhere Enterprise A2019, it is available in 14 languages and has 175 new features for 40 different product capabilities. "This was part of a two-year effort of R&D and extensive feedback from our customers," said Shukla.

But when boiling things down, the main focus was on greatly simplifying the implementation and usage of the software. There is no need to install and configure the system. All you have to do is go to any browser and access it. There are also seamless upgrades.

Another major change with the software was a complete revamp of the user interface. The goal was to reduce the number of clicks and drag and drops to speed up the process.

The bottom line: The benefit of the cloud is likely to be significant and should lead to even better performance with the ROI for an RPA project.

But of course, there are other features and capabilities worth noting. Here's a look:

IQ Bot

From the early days, AI has been a major priority with Automation Anywhere. The technology is continually added to the platform to make it smarter. But there is also a separate system, called IQ Bot. It helps deal with the kinds of processes that deal with enormous amounts of unstructured data. To this end, IQ Bot leverages sophisticated technologies like computer vision, NLP, fuzzy logic, and ML. Furthermore, a customer can use all these without having the need to hire a data scientist or AI expert.

IQ Bot has generated significant results for RPA projects. For instance, the setup is 10 times faster for a business user, without the need to make changes to the workflows. It's also possible to integrate IQ Bot with other AI platforms like IBM Watson.

One of the critical parts of this software is the ability to read and understand business documents, such as insurance claims and invoices. This is not easy to do because of the many complex variations. But IQ Bot has been getting smarter and smarter – and can also work with 18 languages.

To see how IQ Bot operates, let's take an example. Suppose you want better automation for the invoices of your company. These documents can vary widely but with IQ Bot the system can automatically detect fields, such as for the billing address, invoice data, tax, total, and quantity.

You will go to the portal and then proceed with the following workflow:

- Instance Creation: You will fill in the name of the AI component, description, the document type to be analyzed, and the language. Then you will upload a sample invoice and there will be a standard set of fields. Some will be selected and you have the option of adding custom ones.

- Document Analysis: This is the heart of the AI processing, such as classification, semantic analysis, learning from aliases, data extraction, and pattern recognition.

- Bot Training: You will create a bot for the invoice and connect it to the AI component. This is done by using a visual designer. IQ Bot will go on to match fields, and for those that it cannot detect, you can make the connections yourself. You will then test the bot to see if it is working correctly. IQ Bot will also flag any errors, warnings, and validation problems with the fields.

- Production Processing: After the bot is working, you can deploy it. This will take some configuration with the machine that the bot will be used on.

- Document Validation: IQ Bot will provide further testing and will show problems and suggest corrections.

- Progress Monitoring: Through a dashboard, you can keep track of the bot. You'll see metrics on the number of files processed, the straight-through processing (which is the total number of files processed without manual intervention), and the accuracy (the percentage of fields that are accurately identified).

It's still in the early stages but IQ Bot has certainly shown lots of promise. "In five years, IQ Bot will be able to read all documents and determine what's important" said Shukla. "This will be a big improvement for productivity."

Bot Store

The Bot Store is similar to Apple's iTunes or Salesforce.com's AppExchange. That is, you can download bots that have been built by third parties, which can help greatly save time and improve the effectiveness.

The Bot Store is comprehensive, spanning categories like finance/accounting, HR, IT, inventory management, manufacturing, sales, shipping distribution, supply chain management, and support. There is also support for over 30 applications, such as Salesforce.com, SAP, SeviceNow, Microsoft, IBM, and Google.

Launched in 2018, the Bot Store has seen much traction, with 100,000 bot downloads as of mid-October 2019 (this puts it as the largest in the RPA industry). The expectation is that there will be at least 1 million within the following year.

According to Automation Anywhere's own analysis, the use of the Bot Store has resulted in 70% faster development of RPA projects and half the costs.

Automation Anywhere Mobile App

The mobile app is available for both iOS and Google's Android. In terms of the functions, they include the following:

- Start and stop a bot.

- Get access to a dashboard of your bots, such as by monitoring the performance, ROI, activity/usage, and cost savings. There is even a list of the most successful bots.

- Receive custom alerts. For example, you can set an alert that requires attention of an employee if an invoice exceeds $10,000.

- Get access to attended bots that require human intervention. Of course, you can control unattended bots too.

The app shows the data change over time with engaging charts, which show trends in sales, customer feedback, and changes in the operations (see Figure 10-2). What's more, it allows you to add data to the bot. This is done by using the phone's camera to scan a paper document.

Figure 10-2. *Here's the Automation Anywhere mobile app, which delivers real-time access to your RPA dashboard and control of your bots*

Automation Anywhere University

Automation Anywhere has a comprehensive training program, with over 350,000 developers, business analysts, and partners having participated (the goal is to hit one million within a few years). But it is not a purely online system. There are 65 authorized training partners as well as 300+ academic institutions, continuing education programs, and professional associations. For example, there is an alliance with the Certified Professional Accountants organization, which serves the American Institute of Certified Public Accountants (AICPA) and Chartered Institute of Management Accountants (CIMA).

The training is structured based on four main learning trails (develop bots, support operations, define technical architectures, and define use cases). And for each of these, the content is targeted to different roles and skill levels. The languages covered include English, Japanese, Portuguese, Spanish, Korean, German, and French.

There are currently more than 65 courses in the curriculum, all of which are free. But you have to pay a fee for the certification exams.

Validation of Automation Anywhere

Automation Anywhere has received various validations from third-party research firms. For example, in the 2019 report – "The Forrester Wave™: Robotic Process Automation, Q4 2019" – the company was named "Leader" for the highest possible scores in several main categories like bot development, core UI, desktop functions, and market presence (this reflects the number and size of each vendor's enterprise RPA customers and its product revenue).

According to the Forrester report: "Automation Anywhere continues innovation and market expansion [with] one of the largest trained partner ecosystems… more than 800 partners and more than 3,000 customers … [and] increased its global presence to cover all major geographical regions with 40 plus offices…Automation Anywhere remains one of the elite products in the field; clients select its thin client architecture, ease of use, ability to easily connect automations, granular controls for designing robots, and overall low cost of ownership."[14]

To get a sense of the real-world use cases for Automation Anywhere, let's take a case study of one of its customers: Symantec. This is an iconic technology company with 11,000 employees across more than 35 countries.

However, it was struggling with manual processes. True, by implementing an RPA solution, Symantec could see a strong ROI. But this was actually not the priority. For the most part, the company needed to make sure it met rigorous security and compliance requirements.

After evaluating numerous solutions, Symantec found that Automation Anywhere was the right choice. The company then began the implementation in 2017 and also set up a CoE to manage the process. Symantec then used a consulting firm for a term of about six months.

---

[14]www.automationanywhere.com/company/press-room/automation-anywhere-named-a-leader-in-robotic-process-automation-by-independent-research-firm-2019

"Automation Anywhere had enterprise-grade security and worked across different systems seamlessly," said Ravi Konda, who is the senior manager of IT automation and BOTS at Symantec. "The user interface was also intuitive and easy to use."[15]

At first, the focus was on automating the finance department. An illustration of this was the payroll process, which took seven analysts working over a continuous basis for three straight days at the end of every month. But by using Automation Anywhere, the automation resulted in having only two people work on this for just a few hours.

Currently, Symantec has 40 bots deployed that are automating 26 processes.[16] The company is now looking at other areas to apply RPA, such as marketing.

Blue Prism

Blue Prism is the pioneer of the RPA industry, having been launched in 2001. The cofounders were Alastair Bathgate and Dave Moss, who had both worked at a software company that did collections and recoveries. At the time, they realized that current software tools were too complicated to provide wide-scale automation.

Yet the development of the Blue Prism product was not easy. Keep in mind that it was not until 2003 that Blue Prism released the first version of its software, called Automate. But the company had the advantage of having Barclays as an early customer, which helped evolve the system and bolster credibility. The main focus was on rooting out the issues and inefficiencies of collections and recoveries in the call center. From this, Blue Prism would develop an enterprise-grade version of the RPA platform. The main goal was to democratize IT.

---

[15]From the author's interview with Ravi Konda, who is Sr. Manager of IT Automation and BOTS, Symantec, on October 25, 2019.

[16]www.automationanywhere.com/case-study-symantec

By 2016, the company had raised $60.7 million from five investors. Then Blue Prism decided to go public on the London Stock Exchange (under the ticker of PRSM).

Growth has certainly been robust. During the first half of 2019, Blue Prism reported an increase in revenue of 82% to £41.6 million and the addition of 349 new customers, up by about a third from the start of the year. The main growth area was the US market, where revenues nearly doubled.

Yet the company has continued to lose money (during the first half of 2019, there was negative EBITDA of £34 million). Then again, Blue Prism has been investing aggressively in R&D, marketing, and global infrastructure. Consider that the company has over 1,000 employees. The company also serves these key verticals: financial services, telecom, insurance, retail, healthcare and pharmaceuticals, professional services, energy, utilities, public sector, and outsourcing providers.

Unlike many RPA companies that grew out of Silicon Valley, Blue Prism got its start in Warrington England. It happens to be nearby Manchester University, where the first computer was created.

Blue Prism's vision is for connected RPA, which can automate and perform mission-critical processes. And this is not about getting rid of jobs. The company believes that roles will likely change as employees focus more on value-added activities. It's really about liberating the workforce, which should bring more value to the organization.

According to Bathgate, in an interview with Forbes.com: "Connected RPA just completes our vision of being the enterprise play here. A lot of people check their phones in the morning before they even get out of bed. And then you get into the office and you are handcuffed and constrained by the systems and governance of your employers. If you are an innovator, if you are a maker, you are going to get a bit frustrated. That is why lots of little spin-offs and tech start-ups start to nibble away at the very fabric of the profitability of the business. This is particularly acute in FinTech with banking, for example. What Blue Prism tries to connect together is these makers in the organization: connect them to the company strategy and connect them together, because you have people in your organization who have the ideas, but they cannot implement them."[17]

Note   Where did the phrase "robotic process automation" come from? The person who coined this was Pat Geary, who joined Blue Prism in 2008. This was an important development because – until this time – the category was kind of amorphous, referred to mainly as automation software.

A key priority for Blue Prism has been on security. A big part of this is due to the company's roots in the financial services industry. So Blue Prism's approach is to run a secure data center, not rely on a desktop scripting system. By doing this, the company has gotten much traction with large enterprises, especially those in highly regulated industries.

As a testament to the focus on security, Blue Prism was the first RPA company to receive the highest level of Veracode Verified (this validates whether a company's software development processes are secure). The company has also achieved the Verified Continuous designation (this is the top tier analysis for integrated and mature secure practices).[18]

---

[17]www.forbes.com/sites/peterhigh/2019/06/03/blue-prism-ceo-we-developed-the-rpa-platform-digital-exchange-to-free-the-makers-to-innovate/#269477f24535

[18]www.blueprism.com/news/blue-prism-becomes-first-software-vendor-to-achieve-highest-level-of-veracode-verified-accreditation/

All this points to rigorous security practices, such as:

- Using first-party code with static analysis

- Assessing open source software for vulnerabilities

- Integration of security tools in the development process

- Helping with remediation guidance and training

Like the other top RPA vendors, Blue Prism has its own bot store, which is called the Digital Exchange or DX. Some of the most popular apps include bots for machine learning, Microsoft's Azure cloud platform, text analysis, low code/no code, and OCR capabilities. There are also integrations with systems like Salesforce.com, SAP's ERP, Amazon Image Rekognition, and Google Translate.

The process of using DX is straightforward. First, you download the bot to your Blue Prism system. Then it's just a matter of dragging and dropping it onto the process flow. This can greatly speed up the development of bots and also provide much richer functionality.

Blue Prism has been investing more resources in building out its cloud capabilities as well. For example, in the summer of 2019, the company acquired Thoughtonomy for $100 million.[19] Interestingly enough, the firm had never raised venture capital. But it did not matter because it was able to build a strong cloud platform for RPA that leveraged AI, and it worked seamlessly with Microsoft Azure. Over the past three years, the growth in revenues was over 250% (the customer base was primarily with midsize companies).

According to a press release from Blue Prism: "Blue Prism anticipates that cloud-based RPA deployments will in time become increasingly required by enterprise users and, while it currently has numerous cloud-based deployments of its enterprise solution, the development of cloud-based products to further address the market

---

[19]www.blueprism.com/news/blue-prism-agrees-to-acquire-thoughtonomy-to-extend-intelligent-automation-capabilities-in-the-cloud/

requirements is a clear part of its product roadmap. The acquisition of Thoughtonomy, in particular its cloud orchestration tools, resources and skills, will feed into these product initiatives."[20]

To further pursue its innovation, Blue Prism has created the AI Research Labs, which includes a stellar team of PhD researchers, data scientists, and engineers.[21] Some of the goals include:

- Interactive AI: This is where AI systems work with humans, not to replace them.

- Adaptive AI: This involves building AI technology that can change with circumstances over time.

- Trustworthy AI: This is where the models are built with weeding out the bias and other issues that can skew results.

- Explainable AI: This means that the rationales for the AI models will be understandable and transparent.

The first innovation to emerge from the AI Research Labs is Decipher, which leverages OCR and AI.[22] It helps to understand the semi-structured content from invoices (which, by the way, is the most popular request from users of the DX platform). The plan is to expand into other documents like purchase orders, contracts, and even resumes.

---

[20]www.zdnet.com/article/blue-prism-buys-thoughtonomy-for-80-million-eyes-rpa-as-a-service/

[21]www.blueprism.com/news/blue-prism-expands-r-d-capabilities-adding-dedicated-ai-labs-and-outlines-roadmap-for-embedded-ai-capabilities/

[22]www.blueprism.com/news/blue-prism-delivers-new-intelligent-automation-capabilities-to-fuel-connected-rpa-vision/

One of the customers that uses the AI platform is DTE Energy, which has 2.2 million customers and more than 10,000 employees. The company has been able to identify red flags for theft. Basically, only those cases with near certainty are sent to the teams for action. DTE believes that the AI will lead to annual savings of $3 million.

EdgeVerve

Infosys is one of the top consultants for RPA implementations. To leverage on this, the company went on to create its own software, which is run by a wholly owned subsidiary: EdgeVerve Systems.

So why would a consulting firm get into the business of creating its own software products? A big reason is that the company realized that RPA represented a threat to its BPO business. In other words, it's probably better to disrupt yourself than have others do so, right?

I think so.

As for EdgeVerve, it is available for both cloud and on-premise environments. The system also has gotten much traction in verticals like banking, customer service, and credit servicing.

The main offering from EdgeVerve is AssistEdge, which was actually among the four leaders in the 2019 Forrester Wave report (the others included UiPath, Blue Prism, and Automation Anywhere).

The AssistEdge platform has the following:

- RPA: There is a complete offering along with AI capabilities.

- Discover: This is a process mapping tool, which leverages neural networks and machine learning algorithms.

- Engage: This is a specialized module to help improve the performance of the contact center, such as with reduced hold time and faster query resolution.

- Real-Time Expertise Manager: This provides access to subject matter experts.

AssistEdge has about 360 customers and has realized over $2 billion in savings, with more than 50 million transactions per month. One of the customers, a bank in the UAE, used the technology to initially identify more than 21 use cases for automation, such as with past due settlements, budget utilization, and anti-money laundering payment screening. By using AssistEdge, the company was able to realize average 50% FTE savings and 95% process automation.

"We started from AI and then went into RPA," said Seth Siegel, who is the Partner for Artificial Intelligence & Automation at Infosys. "We believe that this has been a big advantage for us. We are looking at higher-order RPA, to find ways to transform businesses. Take a look at the case of intelligent shelves for retail businesses. This is not just about analytics for inventory. With AI, you can go deeper, such as with providing valuable information to the accounts receivable department to get real-time information to obtain better rates on financing."[23]

So is there Infosys consulting required for the RPA software? No, the platform is standalone. This is why Infosys created EdgeVerve as a separate company. Keep in mind that Infosys is one of the largest consulting partners for the UiPath, Automation Anywhere, and Blue Prism.

PEGA

By age seven, Alan Trefler began a life-long passion for chess. By high school, he became a champion in Massachusetts and went on to attain the achievement of a chess master, tying for first place in the World Open Chess Championship with grandmaster Pal Benko.

---

[23]From the author's interview with Seth Siegel, who is the partner for artificial intelligence and automation at Infosys, on December 5, 2019.

But when Alan got into college, he decided to instead leverage his strong analytical skills to pursue a career in software. He would work at companies like Casher Associates and TMI Systems. And then by age 27, he launched his own company, Pegasystems, in 1983.

His big idea? He wanted to develop software to make it easier to develop applications for the enterprise, such as by using visual models that automatically generate code. It was actually a predecessor to low code and caught the attention of large companies, especially banks like Citibank (we'll take a deeper look at low code in Chapter 13).

As for RPA, this part of the business would come much later for Pegasystems (note that the company has since changed its name to PEGA). But it was certainly consistent with the initial vision to help companies transform their businesses with technology without the complexity.

In fact, Alan has some interesting views about RPA. In the second quarter earnings call in 2019, he noted: "Now, RPA can provide value to an organization interested in automating isolated tasks. But the amount of hype and confusion in this market is staggering. We believe that RPA needs to be part of a digital process automation approach, which uses robots in the right context. It's critical to move from thinking robots first, to thinking end-to-end automation where RPA is part of a coherent digital process automation or DPA strategy, not robots being the entire strategy."[24]

PEGA is headquartered in Cambridge, Massachusetts, has 4,600+ employees across 30 global offices, and is publicly traded (NASDAQ:PEGA). The company's platform is called Pega Infinity, which provides an extensive set of enterprise applications that help with:

- Customer engagement: This provides for a system to allow for omni-channel customer contacts and management. The applications help with customer

---

[24]https://seekingalpha.com/article/4283033-pegasystems-inc-pega-ceo-alan-trefler-q2-2019-results-earnings-call-transcript?part=single

acquisition, inbound/outbound marketing, and dealing with paid media channels. There are also modules for sales automation (i.e., helping to manage the whole sales process) and customer service (this goes beyond just tracking data but also helping to anticipate customer demands and issues).

- DPA: This is the part that involves PEGA's RPA offerings (for both attended and unattended bots). There are also a host of prebuilt automations to get started, such as for certain industry verticals and common processes. Since they are built with the low-code system, it is easy to configure them to particular needs and requirements.

- PEGA Cloud: You can host the PEGA applications or the ones you build using low code.

PEGA has infused these products with real-time AI – which it refers to as an always-on "brain" that capitalizes on ML and client data, systems, and touchpoints – and end-to-end automation and robotics. The idea is to go beyond the typical RPA approaches. PEGA Infinity powers the experiences for 1.5 billion consumers.

What if you just want to use the RPA part of PEGA Infinity? This is fine. The platform is designed to allow for the use of any of the modules. For example, as you implement the RPA and achieve your goals, the next step could be to try the other systems that PEGA Infinity provides.

What's more, the RPA modules are divided into two (Figure 10-3 below shows the system):

- PEGA Robotic Desktop Automation: This allows for the creation of attended bots. But it also provides for real-time contextual personalized intelligence for understanding the needs of the customer. Consider that the bot system works across a diverse set of platforms like Windows client applications, .Net, Java, mainframe, web services, and AS400. And PEGA RDA allows for a single-click login as well as helpful dashboards, such as for Start My Day (this shows the order of what applications need to be launched), Customer 360 View, Auto-Note (automatically creates reports on activities), and Shortcuts (these are prebuilt tasks).

- PEGA Robotic Process Automation: This is unattended RPA, which is based on event triggers. There is also the ability to use enterprise features like business process management (BPM) and case management, which allows for even deeper automation.

PEGA got into the RPA business in April 2016 through the acquisition of OpenSpan, a leader in the space (the price tag for the deal was $52 million). The company's main focus at the time was on automating routine tasks for customer service representatives (CSRs) via the desktop. The software was running on more than 200,000 desktops across a host of global enterprises.[25]

OpenSpan got its start in 2004. At the time, tech veteran Francis Carden met with a group of talented Windows programmers who had an intriguing idea to help automate desktop applications. Initially, it sounded mostly about screen scraping. But as Carden talked more to the team, he realized they were describing a much broader vision. He would eventually join the company and make an investment.

Now the reason to focus on the CSR market was that this is where there was a wide assortment of software that made an employee's life quite difficult and inefficient. Because of this, there was simply not enough time to devote to helping customers!

---

[25]www.pega.com/about/news/press-releases/pegasystems-acquires-robotic-automation-software-provider-openspan#sthash.F4yhvv53.dpuf

The main breakthrough with OpenSpan was the development of sophisticated software that would go deep within the Windows layer, such as at the memory level. By doing this, it was possible to allow for the bot and the human operator to work together. It also helped that there was little need for integration. Rather, it was mostly drop and play.

This technology proved to be spot-on and OpenSpan experienced a growth spurt. But for Carden, he was concerned. "After an RPA implementation," he said, "a customer would usually say, 'What's next?' But we didn't have a good answer. This got me wondering that perhaps RPA may have potential issues. So by selling to PEGA, we could offer much more to customers. And I'm still with the company to this day because I believe firmly in this vision."

Figure 10-3.   *This is PEGA's Robot Manager*

Verint

Founded more than 20 years ago, Verint is a developer of software for customer engagement and cyber intelligence. There are more than 10,000 customers across 180 countries and the revenues are over $1.3 billion.

Vertint's RPA platform is part of a broader solution, called the Workforce Optimization Suite. Some of the features include:

- Workforce Management: This allows for accurately forecasting and scheduling employees.

- Automated Quality Management: This ensures that employees are following the right processes, such as by leveraging AI.

- Performance Management: This uses sophisticated data processing to track and improve organizational performance.

- Interaction Recording: This captures communications across the customer journey.

Verint's roots are in the call center market. So yes, this was the initial focus of its RPA platform. But during the past couple years, the company has expanded the capabilities to span other industries.

The RPA system also can handle more complex processes. This is done with the Verint Process Assistant, which allows for real-time guidance and automation wizards, that guide the staff on what to do. In some cases, the app will take over the process.

Then there is the Verint Robotic Process Automation Discovery tool. With this, you can record the activities of the employees at their desktops to map the processes. There is an AI engine that takes the data and provides insights based on the volume, frequency, complexity, and costs. This can then be easily transferred into a bot.

Given its deep experience with the call center, Verint has a strong offering for attended automation. It not only provides assistance to agents but also sends tasks to an unattended bot to finish them. "The Verint solution uniquely offers a robust 'hybrid' automation functionality where the attended and unattended robots can work in tandem moving work back and forth between them," said Ryan Hollenbeck, who is the senior VP of global marketing at Verint.[26]

The RPA solution is also focused on the business user, who may not have much technical experience. "Verint RPA leverages computer vision, among other technologies, to create the robot scripts," said Hollenbeck. "The automations do not require extensive coding or IT support to script and to deploy robots."

WorkFusion

The origin of WorkFusion came from the academic research of cofounders Max Yankelevich and Andrew Volkov at the MIT Computer Science and Artificial Intelligence Lab (CSAIL), which include more than 900 professors, research scientists, postdocs, PhDs, master's students, and undergrads. The organization has an iconic history, with innovation in operating systems, computer vision, and time sharing (which was the forerunner of the Internet).

As for Yankelevich and Volkov, they wanted to leverage statistical quality control and ML to better orchestrate work. They would then create the Process AutoML system that eliminated time-consuming activities for AI, such as data cleansing, the training of models, and testing. The vision was to bring the technology to everyone. For the most part, WorkFusion was the pioneer of cognitive RPA.

---

[26]From the author's interview with Ryan Hollenbeck, who is the senior VP of global marketing at Verint, on October 28, 2019.

The company says that a customer can go live based on the 1-6-12 program:

- 1 Day: A feasibility assessment.

- 6 Weeks: Training.

- 12 Weeks or Less: The solution goes into production.

Here's a case study from a client, which was a large full-service bank in India. The company would receive anywhere from 400 to 500 transaction requests of unstructured data, which was highly manual. There also had to be 45 compliance checks.

As volumes continued to rise, the bank needed to find a better solution. They selected WorkFusion and it took only 60 days to implement a fully digitized process, which extracted the data, classified it, and updated the ERP.

The results[27]:

- The data was in an auditable format.

- Worker productivity increased by 55%.

- A total of 5% of common manual errors were eliminated.

While WorkFusion has worked to simplify its platform, it still requires an organization that has expertise with data science. Keep in mind that building this is expensive, as salaries continue to rise rapidly for technical talent.

In the summer of 2019, WorkFusion announced a strategic partnership with NEC Corporation, a leader in the integration of IT and networking systems. This was to capitalize on the high growth of RPA and AI in Japan. According to the press release: "In April 2019, Japan began implementing work style reform legislation designed to improve the productivity and

---

[27]www.workfusion.com/wp-content/uploads/2018/12/WorkFusion-Case-Studies-Banking-Trade-Finance.pdf

well-being of workers. To adapt to this change and address the needs of a highly-skilled workforce, many organizations are exploring automation and AI to minimize the burden of repetitive work on employees and allow them to focus on value-added projects and initiatives."[28]

Nintex

Founded in 2006 in Melbourne Australia, Nintex initially focused on developing enhancements for Microsoft SharePoint. From this, the company expanded into workflow, digital forms, document automation, RPA, and other automation technologies to cover a wide range of customer requirements. Nintex also has forged partnerships with Salesforce, ServiceMax, and Adobe. As of now, Nintex has more than 8,000 enterprise customers and over 3 million workflow applications in production.

Given this background, the move into RPA was a natural. This was accomplished with an acquisition of EnableSoft in March 2019, which became the developer of Nintex RPA.[29] The software is known for its ease of use and effectiveness with regulated industries.

"We could have partnered with an RPA player but we believed that this technology was strategic for us," said Ryan Duguid, who is the chief of evangelism and advanced technology at Nintex. "We also are focused on providing a comprehensive set of technologies for automation. We think that combining workflow and RPA is the right strategy. One of our customers, for example, uses Nintex to manage over 100,000 discrete processes."[30]

---

[28]www.workfusion.com/news/workfusion-and-nec-partner-to-bring-ai-driven-robotic-process-automation-to-global-markets/

[29]https://news.nintex.com/2019-03-04-Nintex-Acquires-Robotic-Process-Automation-Provider-EnableSoft-Maker-of-Foxtrot-RPA

[30]This is from the author's interview with Ryan Duguid, who is the chief of evangelism and advanced technology at Nintex, on November 25, 2019.

In July 2018, Nintex also acquired Promapp.[31] This was another key piece of the end-to-end strategy as the company allows for optimizing processes with mapping of an organization. Promapp was started in 2002 by two process experts, Richard Holmes and Ivan Seselj, who worked at top consulting firms.

"We think another advantage to Nintex is our business model," said Duguid. "We do not charge by the number of bots on how many machines are used, which can get expensive. We instead have our fees based on each process. Even if there are many machines using this, you only pay the one fee."

Kofax

The origins of Kofax go back to 1985. The founders focused on developing enterprise "capture" software for digital content. But over the years, the company would undergo much change (currently, it is owned by Thoma Bravo, which is a private equity investment firm):

- May 2011: Kofax acquired Atalasoft Inc., an image software company whose primary product was a document imaging toolkit named DotImage.

- December 2011: The company purchased Singularity, a developer of BPM and case management software for public and private cloud platforms.

- Summer of 2013: Kofax acquired Kapow Software, an RPA operator.

---

[31]https://news.nintex.com/2018-07-31-Nintex-Acquires-Process-Management-Leader-Promapp

The result: Kofax has become a full-fledged RPA platform. The company describes this as the "5 capability pillars":

- Process Orchestration: Find ways to improve the customer journey.

- Cognitive Capture: This is about enhancing the processing of documents and electronic data.

- Advanced Analytics: This allows for deep operational insights, such as to reduce risks and eliminate inefficiencies.

- RPA: Use bots to help automate manual processes.

- Mobility and Engagement: This involves the use of secure mobile apps to allow for better collaboration and adoption.

However, Kofax's main strengths are with cognitive capture. Some of the applications include VRS Elite (to automatically enhance the quality of images), Express (to index and export images), Mobile Capture SDK (a system for advanced capture with devices), and AutoStore (for automated and compliant documents capture).

Softomotive

Softomotive is one of the early operators in the RPA industry, having been founded in 2005. At first, the company developed a visual scripting platform (which produced VBScript) for desktop automation.

Softomotive actually bootstrapped its operations and did not raise its first round of institutional funding until September 2018 (this was for $25 million). The company currently has more than 7,000 customers.[32]

---

[32]www.businesswire.com/news/home/20180927005832/en/Softomotive-Raises-25m-Relocates-UK-Base-Global

Softomotive has two main platforms, which handle attended and unattended bots for small to large organizations:

- WinAutomation: This is for desktop automation, which includes an easy-to-use bot designer.

- ProcessRobot: This is geared for larger enterprises, which require high levels of security. To this end, there is seamless integration with Active Directory for user management, authentication, and single sign on (SSO). There is also a centralized store for managing encrypted credentials.

Forrester named Softomotive as a "strong performer" in its Wave report, with the highest scores for innovation/market approach/access to capital, product roadmap, and differentiation.

According the to the analyst's commentary: "[Softomotive's] market positioning is sound, with a business model focused on rapid deployment, good price points, attended use cases, and small-to-midsize businesses, with a growing presence in large enterprises. Softomotive's customers can start small and get fast ROI."[33]

AntWorks

The cofounders of AntWorks, Asheesh Mehra and Govind Sandhu, have a strong background in the BPO industry. Both had been executives at InfoSys Technologies.

AntWorks is a relatively new company, which got its start in 2015. Because of this, the cofounders saw this as an advantage since they could capitalize at cutting-edge technologies like ML and AI as differentiators.

---

[33]www.softomotive.com/softomotive-named-strong-performer-among-robotic-process-automation-vendors-independent-research-firm/

AntWorks describes itself as the "world's first and only Integrated Automation Platform (IAP) powered by fractal science principles and pattern recognition that understands every data type."[34]

The company's product is called ANTstein, which uses low code/ no code for the bot development. Also, the company's reliance on fractal science – instead of Bayesian approaches – seems to work better in business environments.

The result is that ANTstein has a strong capacity for making OCR smarter, in terms of better detection with handwriting, photos, and so on.

In July 2018, AntWorks raised $15 million in a Series A round of funding, led by SBI Investment. The deal also included a strategic relationship for selling into Asian markets.[35]

Intellibot

Intellibot got its start in October 2015. The founders include Alekh Barli (CEO), Kushang Murthy (COO), and Srikanth Vemulapalli (CTO). Both Murthy and Vemulapalli were executives at HSBC, where they managed the automation and testing teams on a global basis. Because of this experience, they were able to see firsthand the gaps in the RPA market.

With Intellibot, the goal was to disrupt the category and redefine it. And a key was extensive product development, as the platform was not launched until February 2018.

---

[34]www.prnewswire.com/news-releases/antworks-is-recognised-as-an-innovator-in-avasants-intelligent-automation-radarview-300929454.html

[35]www.prnewswire.com/news-releases/antworks-announces-series-a-funding-with-strategic-investment-by-sbi-holdings-inc--889224684.html

In terms of the capabilities, they include:

- Cognitive Modelling Platform: This involves leveraging deep learning algorithms to classify and extract unstructured data from many sources like PDFs, images, and JSON files. There are also proprietary engines for natural language processing, predictive analytics, and sentiment analysis.

- Computer Vision: This has helped navigate complex screens, such as with Citrix, legacy software, mainframe systems, and SAP implementations.

- Ease-of-Use: Besides having a visual drag-and-drop approach to bot development, there are also more than 500 components and connectors. Intellibot says that its platform is nearly 100% code free.

- Multitenant Orchestrator: This enables centralized deployment, maintenance, and monitoring of bots.

- Business Model: The enterprise version allows for unlimited bots, which tends to make the system price competitive.

"Intellibot is agnostic of the industry and business functions," said Bharat Madnani, who is the VP of client success at Intellibot. "However, we specialize in BFSI (Banking, financial services and insurance), Manufacturing, IT Ops, Human Resources and Talent Acquisition verticals."[36]

The company is still relatively small, with about 20 customers and a staff of 50. Yet it is growing quickly and is trying to stake out a strong position with cognitive capabilities.

---

[36]From the author's interview with Bharat Madnani, who is the VP of client success at Intellibot, on December 3, 2019.

SAP Contextor

SAP is one of the world's largest software companies, with more than 437,000 customers and revenues of 24.70 billion euros.[37] The company is the pioneer of the ERP market, which handles core functions like payroll, HR, and the financials.

During the past decade, SAP has undergone a major transformation. A big part of this has been an aggressive move to the cloud, such as by making acquisitions for companies like SuccessFactors and Qualtrics.

The company has also made a play for the RPA market. This came in November 2018 when it purchased Contextor SAS. The company begun as an IT services operation that targeted workstation-based projects. But the founders were smart to invest heavily in R&D, which resulted in new technologies – which ultimately became Contextor in 2003.[38]

By joining SAP, the company has certainly benefited from the huge resources – in terms of capital and talent. There has also been significant cross-selling to the existing customer base.

Yet there is a drawback: potential customers may be concerned that Contextor is too SAP focused.

Then again, the software does have clear benefits. Perhaps the most important is the business model, which relies on usage. This could ultimately be much cheaper than the per-bot approach that is typical in the RPA industry.

NICE

NICE stands for Neptune Intelligence Computer Engineering and the company was founded in 1986 by seven former members of the Israeli army. They initially built software for the call center market. But of course,

---

[37]www.sap.com/corporate/en/company/history.html

[38]https://news.sap.com/2018/11/sap-acquires-contextor-robotic-process-automation/

the company has evolved significantly since then, such as with offerings for analytics, AI, and automation. NICE has over 25,000 customers in more than 150 countries, which include 85 of the Fortune 100.

RPA has actually been a natural fit for the company since it has always been focused on improving customer experiences, employee engagement, and operational efficiency. For example, NICE first entered the market about 20 years ago, with its attended automation solution (there is now also an unattended software system).

An interesting part of the RPA suite for NICE is the Automation Finder. This helps to automatically identify potential automation opportunities by leveraging desktop analytics and machine learning (this is a combination of in-house and third-party capabilities).

Another advantage of NICE is that the RPA can lead to other digital transformation opportunities, such as with its customer experience solutions.

Kryon Systems

Based in Israel and founded in 2008, Kryon was created with the vision to help with attended bot automation. This meant that there was an early focus on analytics and AI. For example, the company has since been issued five patents for computer vision and machine learning. Kryon has also looked for ways to broaden its platform, such as with its Process Discovery tool, which helps identify areas to automate. As a testament to the company's technology, there have been many partnerships with companies like Microsoft, Software AG, Amdocs, and EY.

A recent announcement from Kryon was for the launch of AI Booster.[39] It is a platform that operates on Microsoft's Azure Cognitive Services

---

[39]www.kryonsystems.com/kryon-powers-up-its-ai-capability-with-ai-booster-for-even-smarter-rpa/

system. A user can easily create AI automations with drag and drop. Just some of the capabilities include printed and handwritten text recognition in images and the ability to identify fields on business forms.

In February 2018, Kryon raised $40 million in a Series C round, bringing the total amount of investment to $52 million.[40]

Servicetrace

It was back in 2004 that Servicetrace built its RPA system. Located in Germany, the primary focus with its sales efforts has been in Europe.

Servicetrace has taken a comprehensive approach to RPA. In fact, it often takes quite a bit of professional services to understand a customer's processes so as to better optimize things. Note that the focus is on global companies that have complex needs like Vodafone and Airbus (there are about 120 customers).

Servicetrace has also developed tools for other categories including:

- Test Automation (TA): This deals with automating tasks like software upgrades, patches, and updates. Before any actions are taken, Servicetrace will apply a myriad of tests to make sure the changes do not lead to failures.

- Application Performance Monitoring (APM): This is essentially about quality assurance for processes in IT. The software helps make sure the systems are running properly and are not vulnerable to outages.

---

[40]www.kryonsystems.com/kryon-leading-rpa-process-discovery-innovator-raises-40-million/

Conclusion

As seen in the chapter, there is quite a bit to the RPA category! And given the large amounts of investments, there will continue to be new offerings and start-ups. Expect much dynamism, which is certainly great for customers.

For the next chapter, we'll take a look at an emerging area for RPA software – that is, open source providers.

Key Takeaways

- The phrase "robotic process automation" actually was coined by Pat Geary, an employee of Blue Prism. It was important since the category did not have a clear-cut focus or understanding.

- By far the largest RPA vendors include UiPath, Automation Anywhere, and Blue Prism. They are often referred to as the "Big Three." They have massive customer bases, comprehensive platforms, and large amounts of capital.

- This is not to imply that the Big Three will prevail over the long haul. As more capital comes into the RPA space, there will be new innovations, some of which could be quite disruptive. The key areas of focus include AI and process mining. Even open source could be a disruptive force.

- Although, when it comes to AI, the larger RPA companies certainly have an advantage – that is, the significant amount of data to work with.

- Traditionally, RPA software has been on-premise Windows technology. But during the past few years, many of the vendors have been looking to move toward the cloud. This may seem to be late but the reality is that RPA functions are tough to implement with this technology. But companies like Automation Anywhere have been pushing the boundaries, as seen with its latest software release.

- The larger RPA vendors will offer a bot store. This is essentially an iTunes or AppExchange for downloading bots, say, for integrations or adding trade-specific capabilities. This often speeds up the development process. But there are also opportunities for third parties to create new revenue streams.

- Mobile apps have not been common with RPA platforms. But this is starting to change. Automation Anywhere, for example, has been successful with its own mobile apps for iOS and Android.

- To remain competitive and differentiate themselves among the many competitors, RPA vendors have been looking at ways to add more and more functions like process mining, automated documentation, and AI. A key part of this has been through acquisitions, as seen with recent deals from Automation Anywhere, Blue Prism, and UiPath.

- Many of the RPA vendors remain focused on enterprise customers. But this means having sophisticated security systems built in. For companies like UiPath, Blue Prism, and Automation Anywhere, they have been able to make this a part of their development process. They have also worked hard to get third-party validation of their efforts.

- Even though RPA is not highly technical, it still requires training. Many of the vendors have invested heavily in building academies. The Big Three vendors have also been forming relationships with educational institutions.

- While AI remains a top priority for RPA, the technology is still fairly pragmatic. It's often about interpreting forms and transferring data intelligently. All in all, it will be a while until AI can engage in more advanced thinking.

- Some RPA vendors are pure plays. But there are also an assortment of operators that have other portfolios of software, such as NICE, Verint, and PEGA. This can certainly be a benefit as there is seamless integration across the different platforms.

- Another emerging point of differentiation is the business model. Some companies like SAP, Nintex, and Intellibot are using other ways to charge customers, such as by usage.

CHAPTER 11

Open Source RPA

A Look at the Options

In July 2019, IBM announced its largest acquisition ever: the $34 billion deal for Red Hat. A key reason for this was to ramp up the efforts in the cloud market, which has been dominated by Amazon.com, Microsoft, and Google.

But the deal for Red Hat was also a major validation of the open source model.

According to the company's CEO, Jim Whitehurst: "When we talk to customers, their challenges are clear: They need to move faster and differentiate through technology. They want to build more collaborative cultures, and they need solutions that give them the flexibility to build and deploy any app or workload, anywhere. We think open source has become the de facto standard in technology because it enables these solutions. Joining forces with IBM gives Red Hat the opportunity to bring more open source innovation to an even broader range of organizations and will enable us to scale to meet the need for hybrid cloud solutions that deliver true choice and agility."[1]

The irony is that – not long ago – open source was considered not ready for enterprise-grade and mission-critical functions. In June 2001, Microsoft CEO Steve Ballmer famously said: "Linux is a cancer that attaches itself in an intellectual property sense to everything it touches."[2]

---

[1]www.redhat.com/en/about/press-releases/ibm-closes-landmark-acquisition-red-hat-34-billion-defines-open-hybrid-cloud-future

[2]www.theregister.co.uk/2001/06/02/ballmer_linux_is_a_cancer/

© Tom Taulli 2020
T. Taulli, *The Robotic Process Automation Handbook*,
https://doi.org/10.1007/978-1-4842-5729-6_11

Yikes!

But hey, things have changed in a big way for Microsoft too. After all, one of its biggest acquisitions was for GitHub in 2018 for $7.5 billion. The company, of course, is based on the open source model.

OK then, so what about RPA? What have been the open source projects? For the most part, the category is still in the early phases. But as RPA continues to grow, there will likely be more innovation.

So in this chapter, we'll take a deeper look at all this.

What Is Open Source Software?

Richard Stallman was one of the few high school students during the early 1970s that was able to program computers (at the IBM New York Scientific Center). He certainly had a knack for it. When he got into college, he would become a programmer at the MIT Artificial Intelligence Laboratory, where he would stay until 1983. While there, he would make some breakthroughs in AI, such as with intelligent backtracking. He also created the first extensible Emacs text editor.

With his life in academia, he was getting more concerned about how copyrights were restricting access to coding. Because of this, he launched a new movement called free software. This was done by making a new operating system that would rival Unix, which was controlled by AT&T. Stallman called it GNU, which is a clever recursive acronym for "GNU's Not Unix."

But his new approach was not just about giving software away. Stallman allowed anyone to modify and share the code. By doing this, there would be much more innovation than traditional software development, which relied on a limited number of paid programmers. But free software could actually still be sold! Essentially, you could do what you wanted to it so long as you agreed to allow everyone else to use the code you contributed.

But it was not until the early in 1990s that the free software movement would really accelerate. This is when Linus Torvalds, a brilliant Finnish programmer, created the Linux operating system (the key is that he developed a kernel, which was lacking in GNU). It would ultimately become a standard – especially with web servers – and led to the creation of companies like Red Hat.

So then why is the phrase "free software" not used anymore? This was primarily because as this technology became more important for corporate customers, there was a perception problem. Would a CEO want to rely on free software?

Not really.

In 1998, a group of developers came up with the name "open source software."

The Business Model of Open Source?

The concept of open source software does seem counterintuitive. How can something that is given away be at the core of a company? More importantly, how is it that some of these companies are worth billions of dollars?

Keep in mind that open source has different use cases. Take a look at Google. When the company wanted to develop an operating system for mobile devices, it chose the open source model that became Android. At the time, it was a headscratcher for many analysts. Again, how could this be monetized?

For Google, the company realized that the operating system needed to be ubiquitous to have any value. Thus, by making it freely available, this quickly spread the technology, as it became a part of many smartphones from companies like Samsung.

But in the end, Google was able to make lots of money. Part of this was due to the advertising revenues generated from the large install base. But there were also the lucrative fees from the Play App Store. All in all, Android is a platform that has led to billions of dollars in revenues.

Note   According to a lawsuit from Oracle, which claimed that Google illicitly used the Java programming language, Android generated a whopping $31 billion in revenues and $22 billion in profits as of 2015.[3]

In other cases, companies will use open source as a way to build an enterprise software company. An example of this is Elastic. Back in 2004, Shay Banon was on a job search while his wife was taking cooking classes. What to do with his spare time? He developed a search engine, called Compass, for recipes! It turned out to be pretty good and useful for many other use cases. Banon then created another version of the software, which was Elasticsearch and he made it open source. The downloads went through the roof.

In 2012, Banon would team up with Steven Schuurman, Uri Boness, and Simon Willnauer to create the Elastic start-up. After this, there would emerge other applications to bolster the search platform: Logstash (an ingestion tool for sending logs) and Kibana (for data visualization). The founders of the project would eventually join Elastic.

In terms of monetization, Elastic has pursued a hybrid strategy – that is, distributing both open source and proprietary software. In other words, the freely available component allows for much uptake, which then helps to generate interest for the paid versions. Elastic also has fees for support and cloud hosting.

The business model has certainly been spot-on. Elastic, which came public in October 28, 2018, posted revenues of $89.7 million in the fiscal first quarter of 2020, up 58% on a year-over-year basis. The market cap is about $6.3 billion.

---

[3]www.theverge.com/2016/1/21/10810834/android-generated-31-billion-revenue-google-oracle

The Pros and Cons of Open Source Software

Perhaps the biggest advantage for open source software is that it is free. This means a company can try it out and see if it is a fit for its needs. But of course, there are other notable advantages.

One is that there tends to be much innovation, so long as there is a thriving and vibrant community. It's not uncommon for open source versions to be better than commercially available options.

Then there is the benefit of customization, since there is access to the source code. This could ultimately mean a much more effective implementation.

All great, right? Definitely.

Yet open source does have its drawbacks – and they can sometimes be quite serious. Here's a look at some of them:

- Control: Open source software is constantly evolving. But this can be a challenge for enterprise customers. Note that they want a clear-cut roadmap for the platform to allow for better planning. Because of this, companies like Elastic have governance structures where it has the final say on whether a change should be made to the code base.

- Code Bases: An open source project may have two code bases. One is for the community developers and the other is maintained by a for-profit company. However, this can lead to divergences and confusion for customers. There may also be gaps in support.

- UI: Let's face it, developers aren't necessarily good at creating engaging user experiences! And this is actually fine. A developer should focus his or her own efforts on their core strengths, such as on building standout technologies. But this could pose problems with open source software – that is, it may not be particularly easy for typical users. Consider the case with Linux. While it has proven to be quite effective, it has had little traction in consumer and business markets. The fact is that the Windows and the Mac OS platforms still dominate. Then again, these companies realize that great software is more than technology but also requires a focus on design and usability.

- Durability: An open source project may get lots of initial momentum. Yet this can be very difficult to sustain. One reason is that other open source projects may come on the scene that have even better prospects. Regardless of the reasons, this poses big-time risks for users, especially corporate customers. They do not want to invest heavily in an open source platform only to see it evaporate as the community loses interest and goes elsewhere.

- Skills: To use open source software, there usually is a need for extensive setup and configuration. There will also need to be the hosting of the software and ongoing maintenance. No doubt, such activities require having a talented technical staff.

- Backing: Corporate customers often demand that the software vendor provide certain guarantees and commitments, such as with warranties and indemnification. But this is usually not provided by open source projects.

Now for the rest of the chapter, we'll take a look at a few of the top RPA open source projects.

OpenRPA

During the late 1990s, Allan Zimmermann started work in the ISP department at Denmark's largest telco, TDC, as a systems architect. He helped to deploy technologies – like Exchange, NetApp, IIS, Streaming and Lotus Notes – for automation and provisioning. After this stint, he would work at other companies to develop provisioning systems and provide high-end consulting services.

In 2017, he started getting the question: "Can you help us with RPA?" He wasn't too familiar with this segment so he started to learn about it, by testing the Kryon Systems solution. "At the time, I was already creating a workflow engine that combined human workflows and IT/IoT. I tried to create partnerships with several of the big RPA vendors but they all seemed mostly interested in implementation partners. So I then started my own RPA engine, which I now call OpenRPA."[4]

OpenRPA has the core functions you would expect from an RPA system, such as a designer to create bots, a debugger, and orchestration. It is also based on the open source MongoDB database and NodeRED, which allow for handling of huge amounts of data. Actually, this is why OpenRPA has shown promise with categories like IoT.

With OpenRPA, you can save a project locally. This is helpful if the machine is not connected to the Internet and can help deal with strict privacy requirements. With auditing and traceability, and the option to do on-the-fly EAS 265 bit encryption, it becomes easy to comply with GDPR (General Data Protection Regulation) or HIPAA.

There are currently integrations with over 2,000 IT systems that allow for better bot communication. You can also deploy Open RPA on-premise or for different clouds like Azure, AWS, IBM, and Alibaba. "The main difference in our solution is we do 'non-RPA' integration in the

---

[4]From the author's interview with Allan Zimmermann, who is the founder of OpenRAP, on December 8, 2019.

backend, where all others depend on add-ons written specifically for their product," said Zimmermann. "When I started the project, there were no open source solutions based on a workflow designer, and only one had a backend. And it was just a scheduling engine, like most of the commercial products. I created a platform that can support automating everything in a business process, both IT, people and things. We believe that by combining RPA, Workflows, AI and IoT, in an open and secure platform, we can make a positive impact in people's lives."

UI.Vision

Back in 2001, Ann Do and Mathias Roth started iOpus, which developed an app called iMacros. It was the macro recorder tool for web browser forms and it got lots of traction, with over one million active daily users. The founders would ultimately sell the company in 2012 to Ipswitch.

Then in 2016 Do and Roth would launch another start-up: a9t9. "The idea for visual RPA was born out of personal frustration with browser internals," said Roth. "While creating iMacros was fun and rewarding, dealing with the many browser internals and quirks from IE to Firefox to Chrome became increasingly cumbersome. Maybe web developers cannot escape this destiny, but why should users that just want to automate their daily work be dragged into it?"

The company's platform, which is called UI.Vision, is primarily based on the latest innovation in computer vision.

"This is the most powerful approach to automation," said Roth. "Since it operates just like a human does – visually – it can work with any kind of application, web and desktop, and on Mac, Linux and Windows."

For the nonvisual tasks, he recommends the use of established scripting languages like PowerShell or Python. According to him, there is no reason to reinvent the wheel with another scripting system. "PowerShell, Python, and many others are better scripting languages than anything we could create," said Roth. "UI.Vision RPA has a command line

API, so it can be easily combined with these scripting languages. The user can call UI.Vision from the scripting language, and the RPA software can return data back to the scripting language."

Here are some of the use cases: Citrix automations (which can certainly be tricky) and any kind of automation via a remote access tool like TeamViewer, AnyDesk, or RDP.

"Yet other use cases are modern and complex web applications," said Roth. "While web apps can in principle be automated via Selenium-style commands, this becomes often very difficult if the web application gets more complicated. For example, Gmail and Facebook are extremely cumbersome to automate in the classic way."

In terms of the business model, the RPA software is free (it is called UI.Vision RPA Core). But a9t9 does charge fees for add-ons that provide premium features (which is part of the UI.Vision RPA XModules system).

"By design we made sure that now and in the future all Internet access is only done via the open-source core," said Roth. "Our closed source add-on package only runs locally and never 'calls home.' This makes UI.Vision a very secure RPA solution."

Robot Framework

Pekka Klärck is a tester, developer, and independent agile consultant who is focused on test automation. But he wanted to find ways to streamline his efforts. So with his master thesis in 2005, he created the core ideas for the Robot Framework (written in Python), which would be adopted by Nokia Networks. The project would quickly gain adoption, with thousands of developers.

But this platform would go beyond test automation. There would also be an open source system to handle RPA functions, such as with OCR and screen recording. Another key was that the Robot Framework was built for extensibility, allowing for the creation of rich libraries written in Python and Java code. For example, the Jenkins plug-ins provides for orchestration.

Robocorp

Antti Karjalainen has the right blend for the RPA world. While in college, he got two master's degrees in information/service management and automation technology/industrial management. After this, he helped companies with software automation and also led the RPA part of the Robot Framework open source initiative.

Through all this, he knew that the open source model would be the right approach. Thus, in January 2019, he started Robocorp to build an RPA platform from the ground up.

Based in Finland, Karjalainen worked in his garage. "Thank goodness it was insulated," he said.

But to scale the business, he realized he needed funding. "This is critical to create a global ecosystem," said Karjalainen. "Our vision was to make something like GitHub for automation. Instead of sharing code, Robocorp would allow for the sharing of processes."

He would quickly put together an angel round of $500,000 and then by November 2019, he raised $5.6 million. The venture capital firm Benchmark led the round that included Slow Ventures, firstminute Capital, Bret Taylor (who is the president and chief product officer for Salesforce.com), and Rob Bearden, who is the CEO of Docker. As part of the deal, Benchmark partner Peter Fenton became part of the board of directors. He is one of the most successful investors in Silicon Valley, having backed breakout companies like Zuora, New Relic, Elastic, Twitter, and Yelp.

"We are taking a developer-centric focus," said Karjalainen. "And open source is ideal for this. We want something where we can have much more innovation. Traditional RPA is also expensive. A rule-of-thumb is that a bot must replace about a half-year's worth of a person's hours to be profitable. But with the open source model, we should see much more adoption.

We think that RPA can help all businesses. This is something that I call robosourcing."[5]

The core of the Robocorp framework is the Robot Framework (which we covered earlier in this chapter). But there will also be a cloud platform, which will help with the orchestration, security, access management, and so on. These features will likely be provided on a SaaS business model.

In June 2019, Robocorp started a private beta for the software for ten companies from diverse categories like consulting, integration, finance, and logistics. "We are learning a lot and making changes to the software," said Karjalainen. "We are also seeing how RPA is positively impacting these companies. One of our bank customers was able to save 1,369 years of customer waiting time within one month of implementation."

The Robocorp software will be publicly available in the first half of 2020.

Orchestra

While Adam Marcus is a big believer in the power of computers and software, he also realizes that people must be in the loop. "I wanted to understand how you can embed human intelligence in software," he said.[6]

So for his dissertation at MIT's Computer Science and Artificial Intelligence Laboratory (CSAIL), he took on this topic. He would then go on to write a book about this, *Crowdsourced Data Management: Industry and Academic Perspectives.*

But Marcus is not a pure academic. He tested his ideas with an innovative start-up, called Locu, which leveraged machine learning and human participation to determine what every restaurant charges for food. Marcus sold the company to GoDaddy for $70 million in 2013.[7]

---

[5]From the author's interview with Antti Karjalainen, who is the CEO of Robocorp, on December 1, 2019.

[6]From the author's interview with Adam Marcus on December 2, 2019.

[7]http://allthingsd.com/20130819/godaddy-acquires-merchant-finder-startup-locu-for-70-million/

"Through these experiences, we had to build automation software to allow for workers to be more productive," said Marcus. "This would become the foundation for Orchestra, which we launched in 2015."

This is an open source platform that has features like the following:

- Workflows: This is the main project, which provides capabilities for reviews, task management (whether for a person or a machine), and distribution.

- StaffBot: This allows for quick and efficient staffing of projects, which leaves out the extraneous details.

- SanityBot: This is an automated way of monitoring the progress of the project by checking out a Slack channel.

- Time Tracking: This provides an estimate of how long a project or task takes.

TagUI

While working for DBS Bank doing test automation for trading and investment applications in 2016, Ken Soh realized there would be more value applying UI automation techniques to production instead of staging systems. So he left his employer and self-funded the development of TagUI for about a year. After this, he moved the project over to AI Singapore where he became an employee and continued the development during 2018. The government now funds the project.

TagUI definitely has a global dimension, as the platform is based on 21 different languages as inputs to automate workflows. Some of the features include visual automation of web sites and the desktop; a Chrome extension for recording web actions; Python integration for AI and ML; and modules, repositories, data tables, and auditing.

But Soh did not stop here. In 2019, he started on the development of a version of TagUI for the Python programming language.

"I see decentralization as the coming big wave in tech," said Soh. "With this in mind, I developed the Python RPA package to be deployed in a distributed way. And just like the TagUI tool, this Python package is open source and free forever. There is no paywall or registration wall in order to unlock any advanced features."[8]

Besides making it much easier to work with Python and R, the TagUI Python version also has the same automation tools of the previous software as well as OCR automation, keyboard automation, and mouse automation.

"Open source providers have the option of actually combining the best parts of other open source tools to create something stronger, with proper credits of course," said Soh. "The winner for open source RPA may not be the best tool technically, but the one that does the best job of integrating various strengths and minimizing weaknesses of the entire open source RPA ecosystem into a single solution."

Conclusion

Open source still has much to do. But the innovation is showing more momentum. In the years ahead, there will also likely be more funding in the category. This is a trend that we've seen in many other categories of enterprise software.

As for the next chapter, we'll take a look at process mining, which is a technology that is becoming more important for RPA.

Key Takeaways

- Open source is software that is distributed for free and the code base is available for modifications.

---

[8]From the author's interview with Ken Soh on December 13, 2019.

- The roots of open source software go back to the 1980s, when the PC revolution got started. But this category did not see much growth until a decade later with the introduction of Linux. A key driver was the Internet, as the platform became a standard for web servers.

- There are a variety of open source software systems for RPA, such as with OpenRPA, Robocorp, and the Robot Framework. However, the space is still in the nascent stages. Yet there has been much progress and VCs are starting to provide funding.

- Of course, one of the main reasons for the popularity for open source is that it is free. But this software also allows for much innovation because of a global ecosystem of talented developers. Open source also allows for customization, which is a major benefit for larger organizations that have disparate IT systems.

- There are certainly drawbacks to open source. For example, it can be difficult to get a sense of the product roadmap, which can worry large corporate customers. There may also be two code bases, lackluster UIs, and the need for technical talent to deal with the setup, configuration, and hosting of the software.

- While open source software is free, there are business models that allow for monetization. One is to provide consulting and hosting services. An open source company may also have add-ons that require fees.

CHAPTER 12

Process Mining

Using Software to Optimize Processes

As part of a student consultancy in Germany in 2011, a group of college friends – including Alexander Rinke, Bastian Nominacher, and Martin Klenk – begun assisting Bayerischer Rundfunk to improve their customer service department. This began as a typical IT advisory project but the students realized there was an opportunity to create software.

They would leverage something called process mining. This involves using sophisticated Big Data techniques and algorithms to map, monitor, and improve processes – in real-time – by analyzing event logs. In other words, you get a fact-based approach to understanding the processes, such as for areas like P2P (purchase to pay) and O2C (order to cash).

But implementing this was not easy. So the students looked at open source projects to help speed up the development process. There was also a focus on a narrow set of functions and use cases.

And yes, all this worked out extremely well. Rinke, Nominacher, and Klenk would quickly start their own software company, Celonis. The initial capital was 12,500 euros (the minimum in Germany for incorporation).

Despite being small, the company was able to snag enterprise customers like Siemens and Bayer Pharmaceuticals. In fact, Celonis was profitable from day one.

© Tom Taulli 2020
T. Taulli, *The Robotic Process Automation Handbook*,
https://doi.org/10.1007/978-1-4842-5729-6_12

Growth was torrid. By November 2019, Celonis would raise $290 million at a $2.5 billion valuation.[1] The lead investor was Arena Holdings, which was joined by a group that included Ryan Smith, the cofounder and CEO of Qualtrics, and Tooey Courtemanche, the founder and CEO of Procore.

According to Arena Holdings CEO Feroz Dewan: "Celonis is the clear market leader in a category with open-ended potential. It has demonstrated an enviable record of growth and value creation for its customers and partners. Celonis helps companies capitalize on two inexorable trends that cut across geography and industry: the use of data to enable faster, better decision-making and the desire for all businesses to operate at their full potential."

Keep in mind that one of the key drivers for process mining is RPA. Actually, as we've seen with UiPath, some of the RPA vendors are adding their own modules for this.

So in this chapter, we'll take a deeper look at process mining.

Old Way Vs. Process Mining

In Chapters 3 and 4, we looked at how to use process methodologies, like lean and Six Sigma, as well as planning. But this is often manually intensive.

For example, it's common for a company to have workshops where there is lots of work using a white board and putting up Post-it Notes. Or, there may be engagements with consulting firms. Then there will be dashboards, which are hooked to BI and other analytics systems, to help track KPIs.

---

[1]www.celonis.com/press/celonis-raises-approximately-290-million-to-extend-market-leadership-in-process-excellence-software

These are all extremely helpful approaches. But of course, there are drawbacks to consider. First of all, with the workshops and consulting, the activities can be time consuming and also subjective. Let's face it, there is always some politics involved.

Then, regarding the use of dashboards, these are also generally backward-looking and do not offer recommendations. These systems may also not have a holistic view of an organization.

But with process mining, you get a data-driven look at your company's processes. And even better, there is the use of AI and ML to help with insights for improvements.

Backgrounder on Process Mining

In 1939, Carl Adam Petri came up with the graphical way to describe chemical processes – which became known as Petri nets. He was only 13. He would then formalize his ideas in a doctoral dissertation. While it took a while for Petri nets to gain traction, they would eventually become important for parallel and distributed computing as well as workflow management.

These ideas caught the attention of Wil van der Aalst. While at the Eindhoven University of Technology in the early 1990s, he wrote his dissertation about Petri nets.

Yet by the late 1990s – while on sabbatical and then at a one-year stint at the University of Colorado in Boulder – he was running into challenges. "I realized that the models created for process automation, organizational change, simulation, etc., had little to do with reality," he said. "At best, the process models would describe the frequency of executed desirable paths through the process. I was disappointed by the average quality of process models and the impact process models had on reality. As a result, workflow and simulation projects often failed."[2]

Thus was born process mining. He thought that focusing on event data would be a much better approach.

But progress was slow, as van der Aalst had to essentially create a new category of study. "For many years, my research group in Eindhoven was the only group systematically working on this," he said. "At that time, it was not so easy to get data and most researchers did not have a data-science mindset. The people that were working on data mining and early forms of machine learning had no interest in processes."

But he kept pushing forward – and with the past decade or so, process mining started to catch on. Consider that van der Aalst is often referred to as the "godfather of process mining."

He has written more than 220 journal papers and 20 books (author and editor). In fact, he is one of the most cited computer scientists in the world (more than 92,000 references according to Google Scholar). Oh, and he is an advisor to Celonis, as well as other companies like Bright Cape and Fluxicon.

In 2011, van der Aalst published the book *Process Mining: Discovery, Conformance and Enhancement of Business Processes*, and then five years later, he came out with an updated version. In these books, he takes an in-depth look at process mining, including the academic foundations as well as the various tools available.

Van der Aalst has also created an online course for Coursera for process mining, which has had over 100,000 sign-ups (`https://www.coursera.org/learn/process-mining`). It requires about 30 hours to complete.

As should be no surprise, van der Aalst has been instrumental in developing standards and guidelines for the process mining category. To this end, he helped to create the Task Force on Process Mining in 2009

---

[2]From the author's interview with Wil van der Aalst on December 1, 2019.

CHAPTER 12   PROCESS MINING

(this is a part of the Institute of Electrical and Electronic Engineers or IEEE). A few years later, the organization put together the "Process Mining Manifesto."

For the most part, the Task Force helps to organize events, workshops, panels, and conferences to promote process mining. It also encourages research efforts.

How Process Mining Works

Yes, process mining can be very complicated. The academic papers are chock-full of equations and algorithms.

So then, let's take a high-level view of process mining. At the heart of this is an event log, which is the information generated when an application or network takes an action. It could be something like the following:

Event Type: Error

Event Source: Sales

Event ID: 103

Date: 10/12/2019

Time: 9:23:11 PM

User: SALES\YNEE

Computer: OKPKADE

Description: Backup failure.

Process mining will crunch these event logs – which could number in the millions – to do the following:

- Discovery: This looks at the process on an "as is" basis, which is done by using the alpha algorithm or even techniques applied within social networks. Regardless of the approach, you can get a sense of the current mapping of the workflows – and this will be used to create visualizations, which can be truly transformational when understanding and optimizing processes. This can certainly be an eye-opener, as you will quickly get a sense of the bottlenecks and points of friction in the organization.

- Conformance: This involves setting up a model for processes and then allowing the process mining to find deviations. There may also be a decision tree, which looks at a myriad of paths. By using conformance, you should be able to get an even better understanding and optimization with the processes.

- Analytics: Once the data has been collected, the process mining can look at the "to be" processes. This is based on looking at the root causes of the workflows. As a result, you can gain insights on: Is too much time spent on certain activities or partners? Is there enough training in particular areas? Why is one department more effective than another?

It's important to note that process mining is not static. Rather, the technology constantly analyzes event logs. In other words, there are ongoing improvements and insights.

Something else: The perception is that process mining is mostly about working with RPA. But this is really not the case. The technology has proven quite versatile. Here are just some examples:

- Product Development: Process mining has been shown to help reduce the time to market for a new product, provide more consistency with delivery, and decrease duplicate engineering.

- DevOps: With the technology, you can greatly reduce the number of manual changes and the cost to serve – such as by using touchless tickets.

- ERP Migration: Process mining can be critical in improving the success of such initiatives since there needs to be a deep understanding of workflows.

- Internal Audit: By analyzing event logs, it will be easier to identify issues like duplicate payments and missing approvals.

- IoT: This technology throws off huge amounts of event logs. So yes, process mining is likely to be critical in enhancing the analytics.

In terms of best practices with process mining software, this is similar to what we've covered with RPA. Just some of the initial factors to consider include the following:

- Business Case: What are the main goals for the technology? What are the kinds of improvements you want to see? The ROI?

- Pilot: Start with a narrow use case, such as a function within a department. By doing this, you will learn about process mining but also get a sense of the capabilities

of the technology. Along the way, there should be
documentation of the procedures and learnings.

- CoE: This group should also be a part of the process
  mining initiative.

Note   IDC research found that a majority of companies are not
aware of the lost potential due to process weaknesses. Anywhere
from 20% to 30% of revenues can be lost because of this.[3]

Now for the rest of the chapter, we'll take a look at some of the main
players in the process mining market.

Celonis

By far, Celonis is the leading company in the process mining industry, with
more than 600 customers across more than 30 countries. The company has
a solution for both the cloud and on-premise environments. There is also a
free version of the software.

"We think of it as an X-Ray of your company's processes," said
Rinke. "But it is a mistake to say that our technology is a one-time thing.
Celonis provides ongoing monitoring and analysis to help make more
improvements. If all we did was to say what is wrong with a company's
processes, we would have little value."

For example, Celonis has created a myriad of prebuilt modules that
provide insights. These include analytics to show how to get paid faster
and to highlight how to improve on-time delivery within the supply chain.

---

[3]www.celonis.com/process-mining/what-is-process-mining/

Yet Celonis has gone beyond process mining – that is, the software uses task mining. This is a more granular analysis of the actions workers take with software systems. For example, Celonis will use OCR to understand the context of the user actions (such as by identifying topics and cluster activities). This is also leveraged with sophisticated NLP. As a result, based on the data, Celonis may realize that a person is engaging in approving a purchase order and doing research on a particular industry or client. With this, the software can come up with recommendations.

Granted, this may seem kind of creepy. Might this really be just spying on employees? Note that Celonis has built in various safeguards, including filtering based on black and white list applications; allowing the obfuscation of data (this means showing only a snippet); use of anonymization of data, which makes it difficult to find personally identifiable information; and the use of permissions.

To get a sense of how Celonis operates, here's a basic workflow:

- Connect: There is a connection to your ERP (Enterprise Resource Planning) software or other system of record like a CRM. This is how Celonis accesses much of the data.

- Discover: The software will then crunch the data and come up with root causes and opportunities for improving the processes that are presented in engaging visualizations. These will be in the form of key process performance indicators (KPPIs), which can then be connected to your company's KPIs.

- Enhance: With the power of AI, Celonis provides insights for optimizations. The software also has the Machine Learning Workbench. This allows your data science team to use Jupyter notebooks and Python coding to create AI models inside the Celonis platform.

- Monitor: The software will then keep analyzing and learning from your own processes.

Like RPA software, Celonis is fairly easy to use. It's mostly about drag and drop of workflows and configurations. Celonis also has an app store, where you can download prebuilt apps for more than 2,000 implementations (in terms of different industries, software systems, and so on).

All in all, process and task mining are a nice fit for an RPA system. As should be no surprise, Celonis has partnerships with the top vendors like UiPath, Blue Prism, and Automation Anywhere.

Here's what Marc Kinast said about working with Blue Prism (he's the VP of Global Business Development at Celonis): "Business leaders know that successful automation initiatives drive top line value, and bottom line profits, while mitigating risk. We've seen firsthand that organizations often struggle to understand the maturity of their processes, and to decide which processes are standardized enough to benefit from RPA. Process Mining provides a much smoother and more controlled 'process-first' approach to automation, and this approach enables customers to efficiently and effectively transform business operations with powerful analytics and intelligent automation."[4]

One notable case study of Celonis is from one of its partners, SAP. The software giant leveraged the process mining software to help Lockheed Martin streamline its processes.

---

[4]www.prnewswire.com/news-releases/blue-prism-and-celonis-join-forces-to-accelerate-enterprise-automation-initiatives-300710793.html

With Celonis, the company was able to:

- Get visibility across its manufacturing operations (whether manual activities, variant processes, rework, or bottlenecks).

- Access the evaluation of RPA bot performance so as to find ways to improve cycle times and reduce variant processes.

- Product metrics to show the savings.

According to Julia Snook, who is the enterprise integration project engineer at Lockheed Martin: "Celonis Process Mining maps and visualizes processes as they actually occur. From beginning to end, in real time, based on our data, thus ensuring maximum transparency and oversight over our value chain."[5]

ProM

ProM is an open source project for process mining, which is led by van der Aalst. A key advantage to this system is that new research is integrated. In other words, ProM is quite robust and cutting edge. Just some of the whiz-bang features include genetic algorithms and fuzzy logic.

ProM is built on the Java language, which has allowed for a rich ecosystem of tools and plug-ins (there are over 1,500). There are also different variations of the core platform. For example, ProM Lite has basic process mining capabilities. Then there is RapidProM (which is for sophisticated analytics) and Process Mining for Python (PM4Py), a library of process mining modules. "PM4Py provides just a fraction of the functionality of ProM," said van der Aalst. "But it has two advantages over

---

[5]www.sap.com/bin/sapdxc/inm/attachment.2087/pitch-deck.pdf

ProM: scalability and the ability to embed in other systems, for example, as part of a web site."

When it comes to open source process mining tools, ProM is the clear-cut leader. It also has proven to have diverse use cases, such as the following[6]:

- The Dutch National Public Works Department uses ProM to analyze invoices sent from subcontractors and suppliers. The software showed that certain process loops were inefficient because employees were working at remote sites.

- A top manufacturer of semiconductor scanners, ASML, used ProM for its own test processes and found several areas where improvement was needed. The company has also been using process mining to help with less structured processes.

"Open source process mining tools like ProM have been essential for the development of the field," said van der Aalst. "Many techniques and concepts added to ProM can be found in today's commercial process mining tools. This will continue to be the case. This is good because it leads to a transfer of knowledge from academia to industry."

Signavio

When getting his PhD in Business Process Management from the Hasso Plattner Institute, Gero Decker got involved in an open source project that tested what he was learning. He would then go on to work at SAP and McKinsey.

---

[6]www.researchgate.net/publication/221586125_ProM_The_Process_Mining_Toolkit

But he saw that midsize companies were having difficulties improving their processes. They often did not have the talent, such as process experts. The technology was often too expensive and complicated to use.

So he joined several other entrepreneurs to create Signavio (the name comes from the Italian word for signpost). "We were the first company to create a cloud-based platform for collaborative process design," said Decker. "The vision was to bring people together to solve major problems."[7]

Within a year or so, he would abandon the open source project and focus entirely on building the company. Over the years, Signavio would move into other categories like process documentation, target state design, simulation and forecasting, and governance and impact analysis.

And yes, the company entered the market for process mining. This came through a combination of an acquisition and internal development. The first offering was launched in November 2016. Called Signavio Process Intelligence, the system works at high scale across an enterprise and provides real-time insights (Figure 12-1 shows an example of a process map).

---

[7]From the author's interview with Gero Decker, who is the CEO and cofounder of Signavio, in December 4, 2019.

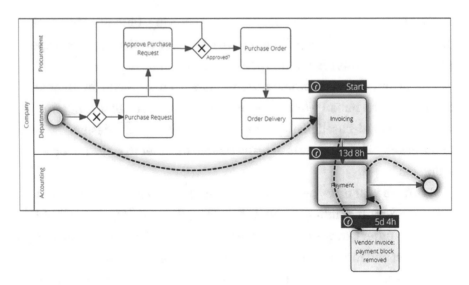

Figure 12-1. *This is a process map created from Signavio's software platform*

"We view process mining as one part of an overall approach," said Decker. "I believe there is too much hype about its abilities. There needs to be a much more comprehensive approach for there to be transformation."

Like Celonis, Signavio is growing at a rapid pace (at about 70% on an annual basis). There are roughly 400 employees, 1,500 customers, and more than one million users.

In July 2019, the company also raised $177 million.[8] The round was led by Apax Digital and included participation from DTCP.

---

[8]www.businesswire.com/news/home/20190711005322/en/Signavio-Raises-177-Million-led-Apax-Digital

Fluxicon

In 2004 to 2005, Anne Rozinat and Christian Günther were in the PhD program for process mining in a group led by van der Aalst. At the time, they were working with large companies like Philips and ASML that provided datasets.

"This was very useful because it helped us to further develop the algorithms to be able to deal with the actual complexity of real-life processes," said Rozinat. "The very first process mining algorithms were namely only able to deal with toy examples. On the other hand, these industry collaborations showed us how people were working with processes in practice, which was very manual, not making use of the data that was collected by their IT systems at all. This inspired us to start Fluxicon. Our goal was to make process mining software that enables companies to analyze their own processes, without having to be process mining researchers themselves."

ProM was gaining steam and both Rozinat and Günther were lead developers for it. For the first few years, they would engage in consulting projects with this tool to learn about what problems needed to be solved. From this came the development of the commercial process mining tool from Fluxicon: Disco (this was in 2012). It was not based on ProM but was built from scratch.

"What is important to realize is that the process mining tools that are available on the market are very different, because they focus on different use cases," said Rozinat. "For example, there are monitoring-centric solutions that follow the classical BI paradigm by connecting to a specific process in a fixed way. Within Disco we do not have a dashboard component on purpose, because most of our customers have their BI tools in place and we see the strength of process mining as an analysis tool."

Fluxicon is an out-of-the box solution that is easy to set up. Actually, the company has a five-minute video that shows how this is done, in terms of the import of the dataset and the automatic construction of the process map (`https://youtu.be/ql1S1wAxJOE?t=5m28s`).

Yet this is deceptively simple. The fact is that there still needs to be domain knowledge to understand the process mining analysis. You need to detect and fix data quality problems as well as have the skills to interpret the results, which can be tricky.

"It is way easier for us to teach companies the methodology for how to apply process mining and to enable them to analyze their own processes rather than to go and do it for them," said Rozinat. "Usually, a two-day training is enough to help them get started. We then further support them from the sidelines."

Fluxicon also has a helpful online book at `https://processminingbook.com`.

ABBYY

ABBYY, which was founded in 1989, is a Digital Intelligence company. It has 1,300 employees and thousands of customers, with over 50 million users.[9] The growth of RPA has been a big driver for ABBYY as most of the vendors license the OCR offerings. It certainly helps that the company has a long history with this technology and can support about 200 languages.

---

[9]`www.abbyy.com/en-us/company/key-facts/`

"OCR is the pillar of automation," said Bill Galusha, who is the director of marketing at the company. "RPA must effectively deal with structured and unstructured content. But OCR is just the first part. There also needs to be machine learning and AI. This is where you can classify and interpret the content, say with invoices or purchase orders. The technology will learn over time how to deal with the nuances and complexities."

Then where does process mining come in? A big part of this has been due to ABBYY's customers, who are often not sure where to start with an automation project.

"We believe process mining is critical for enterprises," said Galusha. "There needs to be effective ways to measure and monitor."

ABBYS's foray into process mining came with the acquisition of TimelinePI, which was struck in August 2019. The founders, Scott Opitz and Alex Elkin, had deep experience with business intelligence and analytics software. They started TimelinePI because they believed that traditional software did not provide an end-to-end visualization of business processes. To pull this off, the software was created to handle huge amounts of data and also leverage cutting-edge AI algorithms.

"There was lots of M&A interest in TimelinePI," said Bruce Orcutt, who is the SVP of marketing at ABBYY. "But the founders went with us because they saw the benefit of combining content with processes."[10]

The company also has a patent pending on its Timeline Analysis approach, which can handle a range of business process types, such as from highly structured to ad hoc.[11]

---

[10]From the author's interview with Bruce Orcutt, who is the SVP of marketing at ABBYY, on December 9, 2019.

[11]www.abbyy.com/en-us/news/abbyy-announces-its-agreement-to-acquire-timelinepi-to-deliver-digital-intelligence-for-enterprise-processes/#sthash.iZUH4Tjc.dpbs

The Future of Process Mining

Even though process mining has led to true innovation in both academia and the business world, it is still in the nascent stages. "It was, and still is, a challenging research topic with many interesting aspects," said van der Aalst. "In the first couple of years, it also became clear to me that the field of process mining is very broad. Next to process discovery, we soon started to work on conformance checking, decision mining, predictions, etc. All of these subfields generate interesting problem statements."

Then what are some of the areas that need to be improved or explored? Where are the opportunities for the future?

Well, here are some thoughts from van der Aalst:

- Process discovery and conformance checking: "These two foundational process mining problems have not yet been solved satisfactorily and there is still a lot of room for improvement. The commercial tools are still using simple Direct Follows Graphs (DFGs) and provide limited support for conformance checking. It is clear for all involved that this is not optimal. Conformance checking is not used as much as expected due to the poor support offered by commercial systems. DFGs have several problems (e.g., underfitting the data and creating non-existing loops whenever activities do not occur at a fixed position in the process). Data extraction also remains an important topic consuming a lot of effort in real-life process mining projects."

- Modelling: "The gap between discovered and hand-made models needs to be bridged. Both need to be integrated and supported in a seamless manner."

- Events: "Traditional process mining techniques assume that each event refers to one case and that each case refers to one process. In reality, this is more complex. There may be different intertwined processes as one event may be related to different cases (convergence) and, for a given case, there may be multiple instances of the same activity within a case (divergence)."

Perhaps one of the biggest hurdles is the talent gap. "Despite the practical relevance of process mining, there are few real process mining experts," said van der Aalst. "Therefore, process-mining training is important to progress the process mining discipline."

Conclusion

Because of its academic roots, process mining remains primarily focused in Europe. But this is likely to change. US investors and RPA vendors are realizing the power of this technology.

But as with any technology, process mining is not a silver bullet. There is still much to be done in terms of the research in areas like conformance and modelling.

Yet even with the limitations, process mining has proven quite effective and versatile. It is also a powerful complement to RPA.

There may also be some confusion with BPM (Business Process Management) and process mining. While both share key elements – especially with the comprehensive nature of managing, tracking, and improving processes – there are clear differences. BPM, for example, often uses low-code approaches and has an extensive set of integrations. As for process mining, this technology relies heavily on using sophisticated ML and AI systems to understand processes. Although, in the future, it would not be a surprise to see more convergence between BPM and process mining. This will also likely be the case with RPA solutions.

As for the next chapter – the last one – we'll take a look at the future of RPA.

Key Takeaways

- Process mining uses sophisticated Big Data approaches and algorithms to map, monitor, and improve processes in real time. This is done by interpreting event logs, which are created by applications like ERPs or CRMs.

- The origins of process mining go back 20 years ago to Wil van der Aalst, one of the world's most cited computer scientists. While at the Eindhoven University of Technology, he researched Petri nets to understand workflows. But he was disappointed in the results. He instead thought a better approach was to analyze event logs.

- In the early days of process mining, there was little interest. But van der Aalst was persistent. He provided much support in the academic community and spearheaded several open source projects. All these efforts would lead to the development of fast-growing companies in the process mining space, such as Celonis.

- Process mining has three main stages: discovery (this looks at processes on an "as is" basis and provides visualizations to get an understanding of the friction and bottlenecks); conformance (this is where a model is established to find deviations, which helps provide for better optimizations); and analytics (this is a look at the "to be" processes like getting insights on what to do next).

- While process mining has been important to the development of RPA, the technology has shown many other use cases. Just a few examples include product development, internal audits, DevOps, and ERP migrations.

CHAPTER 13

Future of RPA

A Look at the Next Few Years

Back in 1965, when computers were still in the formative stages of development, Gordon Moore came up with "Moore's Law" that postulated that the number of transistors on a chip would double about every two years. While it was somewhat off – the timing is usually about 18 months or so – it was still revolutionary. Moore's Law showed the incredible transformative power of computers. Of course, Moore would go on to start one of the world's most iconic companies, Intel.

But there was something else about his law – that is, it is one of the few things in the technology world that has been highly predictable. The pace of change and innovation can be mind-boggling.

Keep in mind that many of the top tech founders and visionaries have been wide off the mark on their predictions. For example, when Apple launched the iPhone in 2007, Microsoft CEO Steve Ballmer had this to say: "There's no chance that the iPhone is going to get any significant market share."[1]

Or how about the time when Digital Equipment Corp. president Ken Olsen said in 1977: "There is no reason anyone would want a computer in their home."[2]

---

[1] https://interestingengineering.com/29-terrible-predictions-about-future-technology

[2] www.inc.com/jessica-stillman/12-hilariously-wrong-tech-predictions.html

© Tom Taulli 2020
T. Taulli, *The Robotic Process Automation Handbook*,
https://doi.org/10.1007/978-1-4842-5729-6_13

So yes, making predictions about the future is a hazardous activity! But then again, it is important to try. There needs to be ongoing debate to help fuel innovation.

In this chapter, we'll take out the crystal ball and make some predictions about RPA.

Consolidation and IPOs

With over 70 RPA vendors on the market – and with perhaps more that will be formed because of the interest of venture capitalists – there will likely be increased consolidation. It will be difficult for many of these players to stand out and get the attention of potential customers.

As seen with operators like UiPath, Automation Anywhere, and Blue Prism, there has already been a pickup in acquisitions. Yet this is likely to accelerate. It certainly helps that the large companies have substantial amounts of cash on their balance sheets.

The acquisitions will be essentially to help expand the RPA technology stack, to move into new industries, and pick up customers. "RPA vendors need to diversify beyond their own niche technologies and core products to grow their position in the industry," said Dr. Gero Decker, who is the CEO of Signavio. "Specifically, vendors will need to better examine the use cases and capabilities that automation can do, therefore expanding the reach and capabilities behind their automation. They also need to make it easier to implement automation, and automate capabilities beyond older, redundant tasks."

There will also likely be some IPOs. Currently, the only pure-play public company is Blue Prism. But in the next few years, we will likely see Automation Anywhere and UiPath pull off IPOs.

This will not just be about raising capital, either. After all, there has been little trouble with this. But being public provides other benefits, including the following:

- Prestige: Not many companies have what it takes to be public. Often, there must be high levels of revenues, a diverse customer base, an extensive product line, and a path to profitability. Being public also raises a company's visibility with the media and analysts.

- Liquidity: Employees may have taken lower salaries in exchange for equity in a company. Because of this, they eventually want to see a return on this – which is much easier when a company is publicly traded.

- Transparency: With many RPA companies private, it is difficult to gauge the industry. What really are the revenues? Is there customer concentration? Any issues with the core infrastructure? For the most part, a public company is legally required to provide full disclosure of material information. This not only makes it easier to evaluate a company's prospects but builds confidence with larger customers.

Microsoft

Historically, Microsoft has been a follower when it comes to new technologies. This was the case, for example, with the GUI-based operating system as well as the Internet browser. It also took several interactions to get these products to a place that was competitive. But when it did, the results were incredible, as seen with Windows 3.0.

Then again, Microsoft has enormous advantages: seemingly unlimited capital, a trusted global brand, a strong ecosystem of partners, and a massive customer base.

So what about RPA? Yes, it's now becoming a priority. At the Ignite conference in November 2019, Microsoft announced its official foray into the market.

The company leveraged its Power Platform – which allows for BI, low code, and workflow management – to create Power Automate. It also uses Selenium, which is an open source application that allows for the recording and automating of web applications.

It comes with more than 275 prebuilt connectors to apps and services. There are also AI capabilities and of course native integrations with Office 365, Dynamics 365, and Azure. While there is some coding needed, it is still high level and there is a drag-and-drop system to create the workflows. Interestingly enough, when it comes to creating alerts and triggers, you can seamlessly use Microsoft Teams.

Power Automate is definitely a solid offering. However, it still lacks enterprise-grade features, such as identity management. But such things will likely be added in the next couple years. If anything, Power Automate will allow Microsoft to learn and iterate on RPA systems. And given the company's deep reservoir of talent and financial resources, it seems like a good bet that the progress will be rapid.

Attended Automation

RPA technology has been mostly about unattended automation. And this should not be surprising. This type of automation is easier to handle. Hey, there is no involvement with people!

But attended automation has much potential for making a huge difference. An employee, for example, can leverage the intelligence of a bot while using his or her own skills to solve problems. Attended automation should lead to much synergy – the best of humans and machines.

The good news is that the technology is starting to get better, especially with improved integration and AI. "We'll see more businesses pivot their attention toward attended automation," said Francis Carden, who is the VP of digital automation and robotics at PEGA. "They will start to recognize how combining bots and humans together can produce optimal outcomes in a faster and more agile way. They will realize that in many use cases, such as in call centers and large front and back offices, attended RPA can be rapidly deployed to many more users for the most common tasks and delivers a much faster and more valuable route to ROI."

Barry Cooper, who is the enterprise group president at NICE, agrees with this thesis. "We're starting to see a shift as enterprises are beginning to understand and embrace the benefits of automation and view it as a way to improve the overall workload and performance of their employees," he said. "This is especially true when providing employees with their own personal robotic assistant. Attended bots are programmed to provide employees with the guidance and assistance they require, in real time directly from their desktops, and automate anything which is repetitive and doesn't require their special skills. Enterprises are realizing that attended automation will have a positive impact on their employee's adoption of automation and ultimately on the service they provide to their customers."

To get a sense of this, consider Automation Anywhere, which has acquired Klevops, a start-up based in Paris.

According to the deal's press release: "Automation Anywhere fast forwards the RPA category to Attended Automation 2.0, where managers can easily orchestrate workstreams across a team of employees and bots, driving a higher level of employee productivity and improved customer experience. This enables customers to automate more processes than ever before, with the same level of central governance, security and analytic capability for which Automation Anywhere has always been known."[3]

---

[3] www.automationanywhere.com/company/press-room/automation-anywhere-acquires-klevops-to-maximize-collaboration-between-humans-and-bots

This approach is likely to see much traction in industries that rely heavily on contact centers, such as banks and telecom companies.

The RPA Career

Even though the emergence of the Internet has resulted in the reduction of certain jobs, there have been new professions created like digital marketing. Something similar is happening with RPA.

"Hiring for RPA skills will explode across all industries and job functions," said Prince Kohli, who is the CTO of Automation Anywhere. "With more than 5,000 RPA jobs in the U.S., there is already incredibly high demand for RPA specialists. Starting salaries will skyrocket."

This will not just be for developers, business analysts, and managers. There will also be growth for those who have experience in verticals, such as IT, BPO, HR, education, and insurance.

But for those companies that implement RPA, there will be a need on how to transition existing workers. This is certainly an opportunity – but will not be easy. Companies must rethink the traditional approaches to talent management. According to research from Leslie Willcocks, Mary Lacity, and John Hindle, change management is the most seriously under-recognized and underfunded component of successful RPA implementations.[4]

"We will see more companies embracing training opportunities to upskill their employees for RPA jobs, opening the doors for new career opportunities and growth," said Barry Cooper, who is the enterprise group president at NICE.

To do this, companies must go beyond just offering training courses. There will need to be a focus on areas like providing interesting opportunities and incentives.

---

[4]www.blueprism.com/uploads/resources/white-papers/KCP_Report_Change_Management_Final.pdf

Scaling RPA

While the early stages of an RPA implementation are generally successful, it usually gets more difficult to scale the technology. What winds up happening is that there is a hodgepodge of bots throughout an organization, which often means not getting the maximum results.

"As more and more organizations adopt RPA, we have yet to see an enterprise-wide adoption of more than 50–100 bots," said Harel Tayeb, who is the CEO of Kryon Systems.[5]

To help improve things, RPA vendors are retooling their systems to help achieve scale, such as by adding AI and process mining. "With the benefits of RPA, corporations are looking for scalability and are struggling in two areas related directly to processes – inability to identify candidates for automation after tackling the 'low hanging fruit' and dealing with bad processes," said Ray LeBlanc, who is the product strategy manager at Verint. "Process mining is the logical solution; however, the solutions are expensive, still take considerable time, and don't truly align with RPA. We are starting to see solutions that are targeted to a smaller sample size, that is, number of users, but also are focused on actual workflow on the desktop, that are focused on discovering processes that are suitable to benefit from RPA."[6]

There will also be a move toward using Business Process Management solutions.

---

[5]From the author's interview with Harel Tayeb, who is the CEO of Kryon Systems, on November 20, 2019.

[6]From the author's interview with Ray LeBlanc, who is the product strategy manager at Verint, on December 10, 2019.

"RPA began as a stand-alone technology but the future of RPA is clearly integration into larger end-to-end process automation platforms and programs," said Michael Beckely, who is the CTO of Appian. "When unified with human workflow, BPM, API integrations, Artificial Intelligence/Machine Learning services, and RPA can be efficiently orchestrated with humans and other systems to maximum benefit. The so-called 'problems' with RPA really become features when viewed within the greater context of an overall Automation project."[7]

He notes that the financial services industry – which was an early adopter of RPA – is already shifting toward a holistic approach. "Whereas RPA was sometimes started as a way to bypass overburdened IT departments, as RPA projects scale, IT is increasingly taking charge," said Beckely. His company has built a system, called the Robotic Workforce Manager, that uses low code to enable the scaling of RPA projects to hundreds and even thousands of bots. For example, Union Bank – one of the largest banks in the Philippines – has used this, along with UiPath, for the unification of its automation initiatives. The company was able to deploy over 400 automated applications in under three months and accelerated processing times by 300%. "Through this process of digital transformation, UBP achieved 97% digitization," said Beckely. "Business executives require real-time visibility into robotic workforce operations so they can analyze bot value by process and department and achieve greater business impact through end-to-end automation. RPA business users need better ways to automate human-in-the-loop activities and self-service controls to start and schedule robotic processes on-demand. With end-to-end process visibility across people and robots, users can confidently tackle much greater workloads than ever before."

[7]From the author's interview with Michael Beckely, who is the CTO of Appian, on December 18, 2019.

Vertical-Specific Companies

In general, most of the RPA vendors have solutions that span multiple industries. Yet the companies still usually have specialties.

Actually, as RPA starts to mature, there will likely be more vertical-based players. Customers will want a more specialized approach, as there will usually be unique use cases.

Just look at Olive, which is targeting the healthcare industry.

The founder and CEO – Sean Lane – actually has an interesting background. He started his career as an intelligence officer in the Air Force working at the National Security Agency. There he had valuable experience dealing with enormous engineering challenges. He even completed five tours of duty in Afghanistan and Iraq.

But when he returned home, he saw another devastation: the impact of the opioid addiction crisis. To do something about this, he looked at how he could leverage his experience with finding insights from data.

When he started learning about the healthcare system, he realized the enterprise systems were highly proprietary and would not talk to each other. The technologies were also unintuitive and often required hiring consultants and outsourcing firms to add new capabilities. This made it particularly difficult to do cutting-edge initiatives like AI.

Because of this, he saw there was an opportunity to bring automation to the industry.

"My focus was on building an AI-powered digital workforce," said Lane. "Instead of building new software to replace disparate systems, we would create the first healthcare-specific digital employee to automate robotic, error-prone workflows, emulating the manual tasks employees had once done – only faster and more accurately. With Olive, I've set out to carve a trillion dollars out of the cost of healthcare while improving the human experience."[8]

---

[8]From the author's interview with Sean Lane, the CEO and founder of Olive, on October 15, 2019.

The name Olive is the "digital person" that provides the automation. She is quick and does her work in a confidential manner, helping with tasks like collections, coding, and credentials. There is also a deep understanding of the complex language of healthcare as well as the EHRs, patient accounting systems, and third-party clearinghouses.

"If something breaks?" said Lane, "Olive proactively fixes it. If a payer portal changes? Olive will adapt her workflow. And if Olive gleans a cost-saving insight from the large amounts of data she's processing? She'll surface it and e-mail her manager about it. This all-in-one approach makes it so automations work for our customers. They don't work for their automations. What's most exciting is that every Olive is able to learn collectively, like a network, so that hospitals never have to solve the same problem twice. I ultimately see Olive's digital workforce as a way to build global awareness – a true internet of healthcare."

The healthcare industry is certainly a prime target for RPA. Note that an estimated $1 trillion in spending in the United States goes to administration. And unfortunately, an employee will engage in mundane tasks like filing folders or moving data across a myriad of systems and interfaces. In light of this, is it any wonder that burnout and boredom are the leading causes of turnover and that attrition in the healthcare industry is so high?

"Imagine what we could accomplish if we eliminated even 1% of the burden of time and cost," said Lane. "The research that could be funded, the treatments that could be developed, the cures that could be found."

Hype Factor

Back in 1995, Gartner analyst Jackie Fenn developed the "Hype Cycle," which caught a lot of attention. The irony is that her framework was, well, hyped quite a bit!

Her premise: Major technologies undergo fairly predictable cycles. In fact, the Hype Cycle has five of them, which include the following:

- The Technology Trigger: There emerges a transformative technology that quickly gets the attention of the media, entrepreneurs, and VCs. There are some prototypes for proof of concepts. But the commercial viability of the technology has yet to be established.

- Peak of Inflated Expectations: At this stage, the technology has shown to be effective and this gins up lots of excitement. This can happen rapidly, especially with the power of social media and the availability of huge amounts of cash to fund new technologies. As should be no surprise, the hype gets to giddy levels. You will hear things like "game changer," "inflection point," and so on.

- Trough of Disillusionment: There begin to be signs that the enthusiasm is evaporating. Some of the companies in the space are failing or not meeting expectations. Stock prices are getting hit. Customers are realizing that the new technology is not generating the promised ROI.

- Slope of Enlightenment: Despite the issues and problems, the technology nonetheless gets more refined. The result is that the impact gets stronger. But the media no longer shows much interest and is focused on other red-hot categories.

- Plateau of Productivity: Here the technology becomes mainstream and there are standards of its use. It is really just a natural part of the business world.

This framework is far from perfect. Consider that Gartner has
had some flubs with its own predictions, such as with BPM and cloud
computing. But the 5-step cycle is still useful. It's a way to allow for a
longer-term perspective and to avoid being too early when it comes to
adopting technologies. According to Gartner: "If there are too many
unanswered questions around the commercial viability of an emerging
technology, it may be better to wait until others have been able to deliver
tangible value."[9]

Then where is RPA on the Gartner Hype Cycle? This is definitely tough
to answer! One of the things to note is that RPA has been around for two
decades and it was not until six or seven years ago that it really started to
accelerate.

But in my opinion, I think the industry is definitely beyond stage two
and may actually be entering the early phases of stage three. The media
attention is intense. The funding has been plentiful. But at the same time,
there are some signs of disillusionment. As noted in this chapter, there are
challenges with scaling RPA as well as combining technologies like AI.

And some of the large players are experiencing growing pains. For
example, in late October, Forbes published an interesting article that
showed that UiPath laid off around 300 to 400 employees or 11% of the
workforce.

This is not to imply that UiPath is in big trouble. Far from it. The
company remains in a strong position – in terms of its product line,
customer base, and financials.

But Wall Street is putting much more emphasis on a path to
profitability. The implosion of WeWork – which lost billions of dollars in
market value and almost went bust – was a wake-up call to many in the
tech world.

---

[9]www.gartner.com/en/research/methodologies/gartner-hype-cycle

This will mean that the strategies will likely be moderating in the coming years. As UiPath CEO Daniel Dines noted in the Forbes article, there will be a need to "balance growth with efficiency."[10]

Software-as-a-Service (SaaS) and Open Source

RPA software is still mostly on-premise. But there will probably be a transition toward a SaaS model. Part of this will be about having a cloud-native platform, which will make it easier for upgrades and data access (this will be critical for AI applications).

"When the software robot becomes a commodity, we will start to see the next wave of adoption in RPA," said Antti Karjalainen, who is the founder and CEO of Robocorp. "SMBs (small and medium size businesses) will want to have access to business process automation but they will not always have the required sophistication to adopt the technology. This gives an opening to a new type of service provider, a robotics-as-a-service (RaaS) operator, that can help SMEs by automating their business routines and maintaining the software robots for them. These RaaS operators can either focus on certain verticals, like car dealers or real estate agents, or they can work as general automation providers in their local area."

But also expect a change in the business model. As RPA scales, the costs can get prohibitive because of the per-bot fees. So customers will be looking for alternatives, say, a subscription approach. Companies like SAP, Intellibot, and Nintex have been offering lower-priced strategies as a way to get traction in the crowded RPA market, for example.

Companies may also look to the open source approach as a way to deal with this as well. But of course, there are other benefits, such as a strong ecosystem of developers who keep innovating the platform.

---

[10]www.forbes.com/sites/alexkonrad/2019/10/24/7-billion-uipath-lays-off-hundreds-in-efficiency-push/#2a03e5cc5f8b

Granted, the landscape for open source software is fragmented. The projects tend to be small in terms of adoption – and the technology is not extensive.

But this is expected to change. As seen with other industries, open source has become ubiquitous.

According to Karjalainen: "The traditional RPA infrastructure hasn't incentivized participation from the developer community to build RPA tools because of the high cost of entry for businesses has essentially limited market-share and profitability. However, as RPA becomes more open-source driven and the user-base grows, developer interest will increase and we'll begin seeing a subset of RPA-exclusive developers that will drive innovative creation of tools. We will identify the 'RPA developer' as a new developer category."

If anything, open source is likely to be critical for the RPA category. "Open source is a central pillar of modern cloud stacks, and if RPA is to have a role in hybrid cloud infrastructure, it must be open source as well," said Phil Simpson, who is the product marketing manager for process automation at Red Hat. "A number of open source RPA projects are available today, but few at this point can compare to proprietary products in the market currently, particularly around providing enterprise-grade support. With that said, the industry is moving toward a more open and flexible model, and I anticipate we'll soon see parity between the two models, giving customers the ultimate in flexibility and choice."[11]

Chatbots

Sometimes there is confusion between chatbots and RPA. While both involve the use of software bots, there are clear differences. A chatbot is a system that uses NLP (natural language processing) to communicate

---

[11]From the author's interview with Phil Simpson, who is the product marketing manager for process automation at Red Hat, on November 5, 2019.

with people. In the consumer world, this would be Siri or Cortana. But of course, many companies are using chatbots for handling customer service.

The market for this technology is growing quickly. According to research from Reports and Data, the spending on chatbots is forecasted to go from $1.17 billion in 2018 to $10.8 billion by 2026, which represents a compound annual growth rate of 30.9%.[12] Some of the biggest drivers include the pervasive use of social media and smartphones.

Chatbots are expected to have a notable impact on the bottom line too. A research study from Juniper Research points out that the cost savings will hit $7.3 billion by 2023, up from only $209 million in 2019.[13]

Then what does RPA have to do with chatbots? There are several use cases. First of all, a chatbot can be used internally as an assistant to help employees with gaining access to information or getting insights. But RPA can also help automate processes that deal with customer interactions.

Look at Aflac. The company, which has been around for more than 55 years, is a top provider of financial protection for health matters. There are over 50 million customers.

No doubt, Aflac has had to find ways to help automate processes. "We have been working with RPA for quite some time," said Keith Farley, who is the VP of US innovation at Aflac. "We have worked with screen scraping and other techniques – even before it was called RPA."[14]

But it was in 2017 that the company implemented two systems from RPA vendors, which involved the help of a consultant. The first bot was for handling wellness claims, which was a quick success. Aflac has since gone on to create 28 more bots. Actually, one of the workers whose processes were partially automated became a manager of the RPA system.

---

[12]www.globenewswire.com/news-release/2019/08/26/1906677/0/en/Chatbot-Market-To-Reach-USD-10-08-Billion-By-2026-Reports-And-Data.html

[13]www.twice.com/retailing/artificial-intelligence-retail-chatbots-idc-spending 33 www.juniperresearch.com/press/press-releases/bank-cost-savings-via-chatbots-to-reach

[14]From the author's interview with Keith Farley, who is the VP of US innovation at Aflac, on December 19, 2019.

"We have reached a point of maturity," said Farley. "We are now looking at the next stage for automation."

A big part of the strategy is the use of chatbots. At first, Aflac leveraged the Facebook Messenger bot service. It was built to address common questions and would route to call center people when the chatbot did not have an answer.

But in the end, it failed and Aflac took it down. But the company did not give up. "We embraced the failure and talked about new approaches," said Farley.

One of the problems was that Facebook Messenger could not identify the user, which made it impossible to provide personalized answers. Next, the database of Q&A information was subpar.

"We thought the solution was to bring people back into the loop," said Farley. "We started using a chatbot system where our employees would answer the questions. By doing this, we were able to create the right kind of data."

This proved to be spot-on. The new chatbot system had a login to identify the users and a much richer database, with the number of question types going from 500 to over 20,000. The result has been a higher level of satisfaction from customers and reduced call center volume during peak times.

"The second iteration was a big success," said Farley. "But we realized that the chatbot was not a replacement for employees. There are certain times when customers want to talk with a person. This is especially the case with our own business, where we deal with wrenching situations with illness and disease."

Artificial Intelligence

AI is likely to be the most important driver for RPA. Based on research from PwC, this technology is expected to add a whopping $15.7 trillion to global GDP by 2030, which is more than the combined output of China and India.

According to the authors of the report: "AI touches almost every aspect of our lives. And it's only just getting started."[15]

Despite this, AI is still limited and narrow. Consider your smartphone that has Alexa or Siri installed on it. While the technology is powerful, the core functionality is really about handling simple commands. It will not be until many years that we will be able to have a free-form conversation with an AI device.

In other words, when thinking about AI and RPA, look for clear-cut business use cases. Interestingly enough, according to a report from PwC, the focus should be on solving "boring" problems. The report highlights that executives are starting to realize this as they run into challenges with implementing projects. The PwC survey shows that – for 2020 – only 4% of companies plan to scale this technology across their organizations.[16] Rather, the priority will be on functions like finance, HR, tax, and compliance. It may be as easy as something like intelligently extracting key information from a form. The survey also shows that 44% of executives see AI as a way to operate more efficiently and 42% believe it will lead to better productivity.

Even before thinking of putting together an AI project or buying a new software platform, it makes more sense to look at your existing systems. Keep in mind that they already likely have AI features built in! For example, companies like Oracle, Salesforce.com, and Microsoft have been investing heavily in revamping their software offerings. So the first step is to try out the new features. This should not only lead to improvements in automation but also provide some understanding of how to work with AI.

Many of the RPA vendors are also aggressively adding AI features (this is often referred to as intelligent automation or IA). But do not rush with these either. Take the time to operationalize the AI within your workflows. What is relevant for your company? What types of AI can be effectively managed?

---

[15]www.pwc.com/gx/en/issues/data-and-analytics/publications/
artificialintelligence-study.html

[16]http://pwc.com/ai2020

Granted, it's a lot of work and there will be the need for extensive planning – but this will be well worth it.

"AI, especially machine learning, has a big role to play in making sense of all the data that will be generated," said Ryan Duguid, who is the chief of evangelism and advanced technology at Nintex. " When you start tracking every mouse click and keystroke, the volume of data is staggering."[17]

Then what are some of the interesting use cases for AI and RPA? Well, there are plenty. Here are just a few, from Tom Wilde, who is the CEO at Indico:[18]

- Corporate E-mail Inboxes: "Most companies have a central inbox that receives lots of e-mails from customers, contractors, suppliers and the like, often with attachments. You can use RPA to detect when a new e-mail arrives with an attachment, then automatically route the e-mail to an intelligent automation tool. Machine learning can then be used to extract the attachment and 'read' it, using OCR and NLP. It can also extract relevant unstructured content such as payment terms, invoice numbers, contractual language and so on. The tool can then normalize the data in an appropriate format and send it to a downstream platform, such as a CRM or ERP tool."

- Contract Management: "Poor contract management can be costly. Businesses might not realize they are owed credits and they may overlook automatic renewal dates, or even fail to send invoices. The automation

---

[17]From the author's interview with Ryan Duguid, who is the chief of evangelism and advanced technology at Nintex, on December 11, 2019.

[18]This is from the author's interview with Tom Wilde, who is the CEO at Indico, on December 9, 2019.

capabilities available in RPA platforms can address many of these issues, but may also be limited in their effectiveness due to the variability of language. For example, provisions and clauses across contracts may use different language but mean the same things. AI can help by understanding context through NLP, and normalizing this information so that the RPA system can automatically alert the right person to address potential issues."

- Invoice Automation: "For invoice processing, RPA can automate data input, reconcile error correction and make binary decisions. But the real challenge is dealing with the many formats different vendors use for their invoices. Using NLP and other machine learning techniques, AI can understand and pull out necessary data from the invoices, normalize it to a structured format, then send it back to the RPA platform for automated data input, error handling, etc."

- Financial Document Analysis: "Financial firms compile lots of data for monthly and quarterly reports. RPA assists by automating data collection from various structured sources. However, once you introduce unstructured PDF documents, RPA is generally ineffective. With the OCR and NLP capabilities of an AI solution, relevant information can be automatically pulled out and converted into a structured format so the RPA tool can deal with it."

- Insurance Claims: "Insurance companies use RPA to automate some aspects of their claims process, such as inputting data from structured sources and ensuring all required fields are filled out. But insurance claims are full of unstructured data, such as photos showing auto damage, PDFs of drivers' licenses, or CT scans for a healthcare insurance claim. AI solutions with machine learning can be used to extract relevant information from these sources, once again adding value to the RPA tool."

With RPA and AI, you may not even have to write up reports anymore! This is possible with a technology called NLG, or natural-language generation.

"Integrating this extends the reach and impact of automation by instantly producing expertly written reports from structured data sets in the form of natural language narratives that are indistinguishable from those authored by business analysts or knowledge workers," said Sharon Daniels, who is the CEO of Arria.[19]

Privacy and Ethics

Governments are getting more aggressive in dealing with regulating the use of data, especially as AI becomes a more powerful force. Europe has enacted the wide-ranging GDPR law and California has introduced the CCPA (California Consumer Privacy Act). These will likely be a framework for new laws in other countries and states.

As RPA gets more sophisticated and goes beyond rules-based approaches, companies will certainly need to be mindful of the new laws – but also be aware of the ethical minefields. A breach can definitely have a deleterious impact on a company, in terms of potential fines and reputational damage.

---

[19]From the author's interview with Sharon Daniels, who is the CEO of Arria, on December 15, 2019.

"The general scope of data privacy laws is to give consumers the right to know how and what type of personally identifiable information (PII) is collected, and the option to take legal action in the event that they should incur damages from bias or data security breaches," said Zachary Jarvinen, who is the Head of Technology Strategy, AI and Analytics at OpenText. "Until now, most organizations have focused their efforts on structured information, but they must also be able to understand what PII is located in textual data documents. Archived data, in particular, is an especially pressing concern for most enterprises. AI-powered solutions will be instrumental in locating sensitive data and managing it through automated workflows. Organizations will also need to establish internal data governance practices to determine who is accountable for data security and enterprise-wide policy, which may include creating teams that blend technical and regulatory expertise."[20]

There should also be steps taken to help deal with the inevitable bias in the data. To this end, models need to make sure the datasets have been cleaned up and are diverse, in terms of backgrounds and characteristics.

Next, companies will need to achieve explainability with the models. Black boxes will ultimately lead to distrust with the outcomes.

Note   Research from DataRobot shows that 59% of companies say they will invest in more sophisticated white box systems (to allow for more explainability), 54% said they will hire internal personnel to manage AI trust, and 48% will use third-party vendors for AI trust.[21]

---

[20]From the author's interview with Zachary Jarvinen, who is the head of Technology Strategy, AI, and Analytics at OpenText, on December 18, 2019.

[21]www.datarobot.com/news/press/datarobot-reports-that-nearly-half-of-ai-professionals-are-very-to-extremely-concerned-about-ai-bias/

Finally, there may be more use of metadata. "This type of information lends itself well to data privacy, and with the correct machine learning and artificial intelligence modeling, can still provide critical information to the C-suite such as lead generation changes, third-party data access, potential breaches and more," said Steve Wood, who is the chief product officer at Boomi.[22]

Conclusion

As AI accelerates, the results should be even better for RPA. We will also see more startups trailbalize new approaches. Just take a look Aisera, which has developed "Conversational RPA." This involves a human-like dialogue interface for business users, providing similar consumer-like application experiences like those of WhatsApp, Instagram, and Snapchat. Conversational AI makes RPA more accessible and extends its reach to a wider audience, unlocking the power of RPA.

In other words, the future certainly looks promising. According to Guy Kirkwood, the chief evangelist at UiPath: "RPA will be the repository for automation. RPA is going to claim its place as a central platform for other enterprise automation tools. Basically, RPA will become the repository for automation in the same way that YouTube is a repository for video content. The centrality of RPA will run parallel with the development of automation code that is more useful and reusable, enabling it to spread even further than it previously has."[23]

---

[22]From the author's interview with Steve Wood, who is the chief product officer at Boomi, on December 5, 2019.

[23]From the author's interview with Guy Kirkwood, who is the chief evangelist at UiPath, on December 5, 2019.

Key Takeaways

- With the large number of RPA vendors, there will likely be increased consolidation in the industry. Some of the larger companies will also pull off IPOs.

- Expect Microsoft to be a major power in the RPA industry. The company's Power Automate platform already is a solid technology.

- Generally, RPA has been about unattended automation. But this will change as technologies like AI become more prevalent.

- RPA career skills will be in hot demand, driving up compensation. Companies will also need to be proactive with training, such as with reskilling.

- Scale has been extremely difficult with RPA. But vendors are looking at ways to deal with this, such as with the use of AI and process mining. What's more, there will be the use of other types of technologies like BPM.

- The Gartner Hype Cycle shows the five stages of a new technology: the technology trigger, peak of inflated expectations, trough of disillusionment, slope of enlightenment, and plateau of productivity. It's not clear where RPA is on this framework – but it seems likely to be beyond the second stage.

- As smaller companies start to use RPA systems, there will be more investment in cloud-native systems and open source projects. Also expect more innovation with the business model.

- A chatbot is not RPA. Rather, it is software that uses NLP to allow for communications with people. But chatbots are important for RPA. For example, they can be used internally as assistants to help employees get information but also to help automate customer interactions.

- AI is likely to have the biggest impact on RPA. The technology will help move beyond rules-based approaches to advanced automation. And as RPA uses more AI, there will be a need to focus on privacy and ethics.

APPENDIX A

RPA Consultants

Atos (www.atos.net/en/) is an IT consulting firm that has over 110,000 employees in 73 countries and annual revenues of more than 11 billion euros. In Europe, the company is a leader with cloud computing, cybersecurity, and high-performance computing.

Regarding the RPA segment, the company has much experience, starting with UiPath implementations back in 2015. Atos also has a major focus on providing advisory services with AI. What's more, the company acquired Syntel, which brought with it the SyntBots RPA platform (it is generally focused on IT operations).

Accenture (www.accenture.com/us-en) is a top professional services company for strategy, digital, and operations, with experience in more than 40 industries. The company has 492,000 employees in more than 120 countries.

No doubt, Accenture can implement RPA at scale. There is also deep experience with analytics and AI.

Capgemini's (www.capgemini.com/us-en/) roots go back more than 50 years. As of now, Capgemini has over 200,000 employees across 50 countries. In 2018, the revenues were 13.2 billion euros.

While the company has experience with multiple RPA vendors, the primary focus has been with UiPath implementations. The company has also been investing heavily in AI/ML with its "Perform AI" framework.

Conduent (www.conduent.com/) is one of the world's largest business process companies (it was spun off from Xerox in December 2016). There are

© Tom Taulli 2020
T. Taulli, *The Robotic Process Automation Handbook*,
https://doi.org/10.1007/978-1-4842-5729-6

about 82,000 employees and the company serves a majority of the Fortune 100 companies and more than 500 government entities.

Conduent has developed its Automated Suite, which helps with RPA and other automation approaches like case management and intelligent automation.

Cognizant (www.cognizant.com/) is ranked 193 on the Fortune 500 and has an extensive set of professional service offerings. In 2019, Forrester gave the firm the highest possible rating for customer and business outcomes as well as partner ecosystem and commercial model criteria for RPA. According to the report: "Cognizant is the RPA turnaround specialist. Cognizant differs from many of its SI peers in that its organizational structure pivots around digital business, giving it a slight advantage in having a value-centric, business-first view of the impact of automation. Its RPA services business has significantly more engagements within line-of-business functions than the typical back-office mix of its peers."[1]

The company has about 2,500 specialists that focus on RPA deployments.

Deloitte (www2.deloitte.com/us/en.html) provides a broad range of services, such as for audit, consulting, and tax across more than 20 industries. The firm has about 80,000 employees and generates close to $20 billion in annual revenues.

Note that Forrester named Deloitte the top consulting firm for RPA (the firm was referred to as the "North Star").[2] Given its overall specialization, it should be no surprise that the company targets finance and accounting. There is also a set of prebuilt tools, including a process discovery system.

DXC Technology (www.dxc.technology/) is one of the early IT consultants (the company is the result of a merger between CSC and the Enterprise Services business of Hewlett Packard Enterprise).

---

[1]https://reprints.forrester.com/#/assets/2/346/RES146255/reports
[2]https://reprints.forrester.com/#/assets/2/346/RES146255/reports

There are more than 6,000 private- and public-sector clients that span 70 countries and there are over 138,000 employees worldwide.

DXC has built a large RPA business, which involves its "robotics-as-a-service" platform (with this, a customer can have their software managed and hosted on the cloud). The department has more than 1,000 automation experts.

In early 2018, the company launched its DXC Agile Process Automation (APA) platform, which involves a blend of RPA with AI.[3] This also involved partnerships with Blue Prism and UiPath.

EY (www.ey.com/en_gl) provides professional services for assurance, tax, and complex transactions. The firm has 260,000 employees and 1 million alumni.

For RPA, EY's largest deployment is its own! The firm also has focused primarily on Blue Prism for its implementations and there is a growing business with SAP. Furthermore, EY has built EYSight, which helps with assessing a company's processes.

In 2018, HFS Research named the firm as the top player for RPA services (the survey included 29 vendors). According to Elena Christopher, who is the research VP of HFS Research: "Though RPA has emerged as a powerful change agent for enterprises, it is still a nascent market and service providers play an important role throughout the life cycle of planning, implementation, management, operation and optimization. EY teams have leveraged RPA to support and scale end-to-end process transformation for clients and exhibited a strong mix of service execution excellence, applied innovation and vision, and verified customer satisfaction to rise to the top of our RPA services study."[4]

---

[3]www.dxc.technology/newsroom/press_releases/143719-dxc_technology_launches_agile_process_automation_to_transform_business_processes_with_data_discovery_robotics
[4]www.ey.com/en_gl/news/2018/11/ey-ranked-one-in-robotic-process-automation-services-by-h-f-s-research

Genpact (www.genpact.com/) provides both consulting and BPO services. There are more than 87,000 employees and revenues of $3 billion.

Interestingly enough, with BPO, the company already has access to the various major processes in a company – say, for finance or IT – and has continued to look at ways to optimize systems, say, with lean and Six Sigma. Because of this, RPA services are a natural extension. All this is part of a "lift-shift-transform" model. Genpact has also made a string of acquisitions, such as for PMNsoft, TandemSeven, and RAGE Frameworks, to bolster its automation efforts.

Hexaware (hexaware.com) is a provider of IT and BPO consulting services (its moto is "Automate Everything, Cloudify Everything, Transform Customer Experiences"). The firm has 33 global offices, more than 18,000 employees, and annual revenues of $677.7 million.

Hexaware has a detailed business process framework that helps deal with the complexity, transaction volumes, and legal/regulatory requirements. After this, the company helps put together a strategy for a comprehensive implementation. Hexaware has partnerships with the following RPA vendors: Automation Anywhere, WorkFusion, UiPath, Blue Prism, and PEGA.

HCL Technologies (www.hcltech.com/), which has been around for more than four decades, has developed the Mode 1-2-3 strategy for providing IT services to clients to help with trends like the cloud, analytics, IoT, and automation. The framework covers core services and next-generation services, products, and platforms.

HCL has more than 147,000 employees (which the company calls "ideapreneurs") that work across 44 countries. It also serves half of the Fortune 500 and 650 of the Global 2000.

With RPA, the company has full-on services and has deep experience with verticals like banking, capital markets, public services, healthcare, telecom, consumer services, and manufacturing.

Infosys (www.infosys.com) was founded in 1981 by seven engineers in India (the start-up capital was a mere $250). But the company would ultimately become one of the largest IT consulting firms. Today it has 229,000 employees (10,000 are based in the United States) and revenues of $12.4 billion. In 2019, Forbes named Infosys No. 3 for the World's Best Regarded Companies, based on factors like "trustworthiness, honesty, social conduct, fairness to its employees and the performance of its products and services."[5]

The company's RPA practice has more than 1,000 consultants and developers that use a proprietary platform called AssistEdge, which is used by 360 enterprise customers. In fact, it has realized $2 billion in savings and currently process 50+ million transactions every month.[6]

Infosys also has support for RPA providers like Automation Anywhere, Blue Prism, UiPath, and WorkFusion.

KPMG (https://home.kpmg/xx/en/home.html) operates in 147 countries and has 219,000 employees. Regarding RPA, the company has emphasized process mining. Consider that KPMG has a major relationship with Celonis, which is the top software provider in the category. The firm has also developed the RPA-TOM (Target Operating Model) approach, which is a six-step process for RPA implementations.

Mphasis (www.mphasis.com) operates with a system it created called Front2Back, which leverages cloud and cognitive technologies to allow for more personalized digital experiences. There is also its system that helps with the transformation of legacy IT environments.

PwC (www.pwc.com/) was founded in London over 100 years ago as an accounting services firm. But since then, PwC has expanded significantly, with over 276,000 employees in 157 countries. Just some of its services include assurance, tax, cybersecurity, HR, and forensics.

What about RPA? It's certainly a growing part of PwC's portfolio. The firm helps with the initial implementation and ongoing management,

---

[5]www.infosys.com/newsroom/press-releases/Pages/worlds-best-regarded-companies2019.aspx

[6]www.edgeverve.com/assistedge/

such as with the development of a CoE. There has also been success in specialized areas for RPA like the audit function.

TCS (www.tcs.com/) is an IT services and business solutions powerhouse, with more than 450,000 employees in 46 countries and consolidated revenues of $20.9 billion. Note that the firm has a long history with providing advisory services for automation, such as with its outsourcing operations. And yes, RPA is a major part of this. Among the large firms, TCS is known to have relatively lower costs.

APPENDIX B

RPA Resources

RPA web sites:

- Horse for Sources Blog: www.horsesforsources.com

- RPA Today: www.rpatoday.net

- CIO.com: www.cio.com/

- TechRepublic: www.techrepublic.com/

- The Enterprisers Project: https://enterprisersproject.com

- IEEE: www.ieee.org/

Software review sites that cover RPA:

- www.g2.com/categories/robotic-process-automation-rpa

- www.gartner.com/reviews/market/robotic-process-automation-software

- www.trustradius.com/robotic-process-automation-rpa

RPA influencers:

- Alastair Bathgate, Blue Prism CEO: @AlastairPRSM

- Pat Geary, evangelist at Blue Prism: @EvangelistBlue

- Mihir Shukla, Automation Anywhere CEO: @MihirAndNow

© Tom Taulli 2020
T. Taulli, *The Robotic Process Automation Handbook*,
https://doi.org/10.1007/978-1-4842-5729-6

- Gary Conway, chief evangelist at Automation Anywhere: @GaryGConway

- Daniel Dines, UiPath CEO: @danieldines

- Guy Kirkwood, chief evangelist at UiPath: @guykirkwood

- Kirk Borne, principal data scientist and executive advisor at Booz Allen Hamilton: @KirkDBorne

- Frank Casale, the founder of The Outsourcing Institute: @FrankCasale

- Mike Quindazzi, the digital alliances sales leader at PWC: @MikeQuindazzi

- J. P. Gownder, the VP tech analyst at Forrester: @jgownder

Other Books:

- Service Automation, Robots, and the Future of Work (2016), Willcocks and Lacity

- Robotic Process Automation and Risk Mitigation: The Definitive Guide (2017), Lacity and Willcocks

- Robotic Process Automation and Cognitive Automation: The Next Phase (2018), Lacity and Willcocks

- Becoming Strategic with Robotic Process Automation (2019), Willcocks, Hindle, and Lacity

APPENDIX C

Glossary

AI (artificial intelligence): Involves computers that are capable of learning from experience. This is generally done by processing data with sophisticated algorithms. AI is a broad category that involves subsets like machine learning, deep learning, and natural language processing (NLP).

API (application programming interface): This is software that connects two applications. It is actually a more complex form of automation vs. RPA because it requires the assistance of developers.

Assisted RPA (or attended RPA): This involves automation of processes that still require human collaboration. This is the first generation of RPA technology.

Autonomous RPA: This is essentially a blend of attended and unattended RPA.

Big Data: Technology that allows for processing enormous amounts of data. Data is often described as having the three V's: volume, variety, and velocity.

Black box testing: This is a form of testing software where the tester does not know the internal code but instead looks at the input and output to see if there are any problems.

Boolean: This is a data type that is either true or false.

Bot: This is at the heart of an RPA system. A bot automates a set of instructions, such as moving data and pressing buttons on an app.

Bot Store: This is a marketplace, similar to iTunes or Google Play, where you can download bots into your platform.

© Tom Taulli 2020
T. Taulli, *The Robotic Process Automation Handbook*,
https://doi.org/10.1007/978-1-4842-5729-6

Breakpoint: This is used for debugging code. That is, there is a termination of the program at certain points allowing for the coder to see what is happening.

Business analyst: This person manages the duties between the company's SMEs (subject matter experts), RPA supervisors, and developers.

Business continuity plan: This is a document that sets forth the goals and actions to be taken when things go wrong with a bot or an RPA implementation.

Business process management (BPM): This type of software has been around since the 1980s. Generally, it is much more comprehensive than RPA. Although, because of this, the implementation of BPM can be more costly and time-consuming. The technology also requires much effort from an organization to maintain.

Business process outsourcing (BPO): This is an organization that manages outsourced services, such as for back-office operations.

Categorical data: Data that lacks numerical meaning but instead has textual meaning, say, with describing race or gender.

Center of excellence (CoE): This is a group of people who implement, deploy, and manage an RPA system. It can be small (say, a few people) and may involve persons outside of the company, such as consultants.

Chatbot: AI software that allows for the communication with people. Examples include Siri, Alexa, and Cortana.

Cloud: This is where you can store and manage data and applications from the Internet, which is known as the public cloud. But there is also the private cloud (where access to the servers are restricted for security purposes) and the hybrid cloud (a combination between private and public clouds).

CoE: See Center of excellence.

Cognitive RPA: This is where AI is used with RPA technologies. One of the common approaches is NLP, which interprets voice commands and written content. By using this type of technology, the RPA system will learn over time (such as how to understand invoices and other business documents).

CRISP-DM Process: Developed by academics, consultants, and experts, this is a 7-step process for managing data in a project.

Customer relationship management (CRM) software: This helps to manage a company's relationships and interactions with contacts, leads, and customers.

Database: Software that allows for storing and retrieving information, which is critical for any type of application.

Data type: The kind of variable used in a programming language, such as a Boolean, integer, string, or floating point number. Note that data types are often used in building bots.

Deep learning: This is a subset of AI that involves the use of sophisticated neural networks. During the past decade, much of the innovation in the AI field has come from deep learning research.

Enterprise resource planning (ERP) software: This helps to manage the core functions of a company, such as financials, human resources, and the supply chain.

ETL (extraction, transformation, and load): A kind of data integration that is typically used in a data warehouse.

Explainability: This is where you try to understand the underlying rationale of a deep learning model.

Flowchart: This is a visual representation of the steps in a process. This is also core to an RPA system. A flowchart may also be referred to as a sequence.

Gartner Hype Cycle: A framework for describing the cycles of a new technology. It involves five stages: the technology trigger, peak of inflated expectations, trough of disillusionment, slope of enlightenment, and plateau of productivity.

GPUs (graphics processing units): These are semiconductors that are for high-speed video games because of the ability to process large amounts of data quickly. But GPUs have also proven to be adept at handling AI applications.

Grey box testing: This is testing of software that includes both white box and black box testing. Essentially, it is a very comprehensive approach.

Hadoop: Open source software that helps with managing Big Data, say, by making it possible to create sophisticated data warehouses.

Hidden layers: The different levels of analysis in a deep learning model.

Hybrid cloud: This is when a company uses both a private and public cloud.

Integer: This is a data type that holds a number without decimal points.

Intelligent automation (IA): This is where AI is used to enhance an RPA platform.

Key recording: This is a standard feature in an RPA system, which captures a person's keyboard/mouse movements. These are then embedded in a bot.

Lean: This is a process methodology whose origins go back to Toyota during the 1950s. The focus is on constantly finding ways to improve a system.

Lean Six Sigma: Involves a combination of the two process methodologies. With lean, there is a focus on the elimination of waste and other inefficiencies and Six Sigma helps with data and statistics. A typical approach is to first use lean and then go to Six Sigma.

Log message: This is a comment you can make within a step in an RPA process, which will show up in the orchestrator. Log messages can be very helpful with documenting and debugging.

Loop: A common structure in programming languages that allows for repeating a set of instructions until a condition is met.

Low code: Allows the development of software applications with little hand coding, which often means quicker development.

Macro: This is a series of steps or actions that are automated for a particular application, such as Excel. This is not really RPA, though. RPA instead is much more comprehensive in terms of what can be automated.

Machine learning: This is a subcategory of AI, which uses traditional statistical techniques to help with predictions and insights from processing data.

Metadata: This is data about data – that is, descriptions. For example, a music file can have metadata like the size, length, date of upload, comments, genre, and artist.

Natural language processing (NLP): A subcategory of AI that involves the use of software to understand and manipulate language. Common uses of NLP include Siri and Alexa.

No code: Allows the for the development of software applications by using simple approaches like drag and drop.

Normal distribution: A plot of data that looks like a bell and the midpoint is the mean. A normal distribution (also known as a bell curve) has been shown to be common in nature, such as with weights and heights.

NoSQL system: This is a next-generation database. The information is based on a document model in order to allow for more flexibility with analysis as well as the handling of structured and unstructured data.

OCR (optical character recognition): This is a document scanner that recognizes text, such as from images or even handwriting.

On-premise software: Where a company installs and maintains its own technology within its data center, which allows for more control, security, and privacy. But it can be costly and difficult to customize. This is why more companies have been moving to the cloud.

Open source: This is software that is freely available. Anyone can enhance the code so long as they agree to provide this for everyone without a fee. Open source is also becoming more of a factor for RPA.

Orchestrator: This is a platform that helps to manage the bots, such as with scheduling, tracking, and termination.

Private cloud: This is a cloud system in which a company has its own data center.

Process mining: Sophisticated software – which leverages Big Data and algorithms – to map, monitor, and improve company processes. This is becoming an increasingly important part of RPA implementations.

Public cloud: This is when a company will store and maintain its software and databases on another cloud platform, such as Amazon Web services or Azure.

Python: A computer language that has become the standard in developing AI models.

Reinforcement learning: This is a type of algorithm that is trained by rewarding accurate predictions and punishing for those that are not.

Relational database: A database, whose roots go back to the 1970s, that creates relationships among tables of data and has a scripting language, called SQL. Relational databases are the most common within corporate environments.

Robotic process automation (RPA): A software platform that allows you to automate specific business tasks, such as moving data and making keyboard inputs. Often, this is done with applications like a CRM or ERP.

RPA champion: This is the evangelist for the project. He or she will focus on such things as creating videos, workshops, and blog posts.

RPA developer: This person designs, deploys, and monitors the bots.

RPA infrastructure engineer: This person is responsible for managing the server installation.

RPA solution architect: This is someone who assists with the early stages of a project, such as with the design of the core technology foundation.

RPA sponsor: This is a person from the business side of the organization that provides support for the implementation.

RPA supervisor: This person manages the team players of the project.

Scope: This is the part of a program where a variable can be used.

Screen scraping: Involves the transferring of data from one application to another. This is one of the features of an RPA system.

Script: This is a set of programming instructions, such as for the creation of a bot.

Semi-structured data: This is a blend of structured and unstructured data, which usually includes internal tags for categorization.

Sequence: This is a visual of the different steps of a bot in an RPA process. It is usually from top to down and has different icons to note things like decisions and so on. A sequence may also be referred to as a flowchart.

Six Sigma: This is a process methodology that was developed in the 1980s by Motorola. Six Sigma relies heavily on statistical techniques to help reduce defects in a system.

SME: See Subject matter expert.

Standard deviation: This measures the average distance from the mean, which gives a sense of the variation in the data.

Strong AI: This is true AI, in which a machine is able to engage in humanlike abilities like open-ended discussions. However, current technology is far from achieving this.

Structured data: This is data that has a certain format (social security number, address, point of sale information) that can be stored in a relational database or spreadsheet.

Studio designer: This is the part of an RPA software platform where you can design bots. Often this involves light code and drag and drop.

Subject matter expert (SME): This is a person within an organization that has expertise with certain processes.

Supervised learning: This is a type of algorithm that analyzes labeled data. Supervised learning is the most common in AI.

Test data: After a model is created, you use this type of data to evaluate the results.

Training data: This is the data used for creating a model, such as for machine learning or deep learning.

Unassisted RPA (or unattended RPA): Where the software completely automates a process or task.

Unstructured data: This is data that is unformatted, such as images, videos, and audio files.

Unsupervised learning: This is a type of algorithm that uses unlabeled learning. This process often involves looking for clustering of the data.

Variable: This is a container that holds data that can be manipulated in computer code.

Virtual assistant: An AI device that helps a person with his or her daily activities.

Weak AI: This is where AI is used for a particular use case, such as with Amazon.com's Alexa.

White box testing: This is testing of a software's source code.

Index

A

ABBYY, 288–289
Algorithms
    reinforcement learning, 210–211
    semi-supervised
        learning, 211–212
    supervised learning, 209, 210
    unsupervised learning, 210
American Institute of Certified
        Public Accountants
        (AICPA), 231
AntWorks, 176, 185, 188, 250, 251
Application Performance
        Monitoring (APM), 255
Application programming
        interfaces (API), 37–38, 325
Artificial general
        intelligence (AGI), 40
Artificial intelligence (AI), 5, 325
    components, 40
    conquer disease, 38
    contract management, 310
    corporate e-mail inboxes, 310
    deep learning, 39
    digital customer service
        platform, 20
    insurance claims, 311

    invoice automation, 311
    issues
        bias, 41
        black box, 42
        brain, 43
        causation, 41
        common sense, 41
        comprehension, 42
        conceptual thinking, 42
        static, 42
    machine learning, 39
    NLP, 40
    structured data, 43
    unstructured data, 43
AssistEdge platform, 238
Attended automation
    Automation Anywhere, 297
    benefits, 297
    integration and AI, 297
    RPA career, 298–299
    scaling RPA, 299–300
Attended bots, 108, 111, 218, 219,
        230, 241, 245, 250, 254, 297
Attended RPA, 6, 15
Automate, 233
    alerts, 88
    customer cancellation process, 87

© Tom Taulli 2020
T. Taulli, *The Robotic Process Automation Handbook*,
https://doi.org/10.1007/978-1-4842-5729-6

S

Printed in the United States
By Bookmasters